28

W9-BZB-629

DISCARD
The Way of the World

SEP 5   2008

ALSO BY RON SUSKIND

The One Percent Doctrine
The Price of Loyalty
A Hope in the Unseen

SEP S 2008

973.931
S964W
Large
Print

The Way of the World

A Story of Truth and Hope in an Age of Extremism

Ron Suskind

An Imprint of HarperCollins*Publishers*

MORSE INSTITUTE LIBRARY
14 East Central Street
Natick, MA 01760

THE WAY OF THE WORLD. Copyright © 2008 by Ron Suskind. All rights re-served. Printed in the United States of America. No part of this book may be used or reproduced in any manner whatsoever without written permis-sion except in the case of brief quotations embodied in critical articles and reviews. For information address HarperCollins Publishers, 10 East 53rd Street, New York, NY 10022.

HarperCollins books may be purchased for educational, business, or sales promotional use. For information please write: Special Markets Department, HarperCollins Publishers, 10 East 53rd Street, New York, NY 10022.

FIRST HARPERLUXE EDITION

HarperLuxe™ is a trademark of HarperCollins Publishers

Library of Congress Cataloging-in-Publication Data is available upon re-quest.

ISBN: 978-0-06-156283-9

08 09 10 11 12   ID/RRD   10 9 8 7 6 5 4 3 2 1

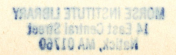
MORSE INSTITUTE LIBRARY
14 East Central Street
Natick, MA 01760

To Cornelia, who inspires hope

As a general rule, people, even the wicked, are much more naïve and simple-hearted than we suppose. And we ourselves are, too.

—FYODOR DOSTOYEVSKY,
   The Brothers Karamazov

Contents

Act III: The Human Solution

The Way of the World

Prologue:
Border Crossing

FROM THE DAWN OF TIME, human beings have been attentive to signs of distinction—the approach of a tribe with a different manner or dress, posture or skin color. The swift sizing-up of friend or foe, and acting upon it—upon suspicion—was often a matter of survival. Those faculties became finely tuned over thousands of years.

Now, in a world of vivid, colliding images and technology's bequest of awesomely powerful weapons, we struggle to leap forward, to reshape instinct enough to reach across the divides of us and them, peak and valley. And to do it in time.

That shared effort is, at the very least, a starting point for a working definition of "hearts and minds struggle," that smooth, slippery phrase on the lips of people across the world. Its definitions are often self-interested and oddly narrow, but they nearly always rest on a fundamental two-part question: Can disparate people ever truly understand one another, and is such understanding necessary for them to coexist? There's considerable dispute over the matter. Some knowledgeable observers say that bringing diverse peoples together mostly serves to exacerbate distinctions and fuel divisiveness, something we can little afford in an era of such unleashed destructive capability. They point to countless bitter conflicts along borders, and within them, and recommend tall fences. Others contend that the world is steadily becoming borderless and blended, and that such conflict—the friction caused by the conjunction of opposites—must be endured, and mastered, on the way to discovering shared interest and common purpose.

One rare area of agreement? That the answers—whether proof of insurmountable divides or of indelible human bonds—are found by walking in the shoes of the "other."

The *other*? Could be anyone, really, from the person who seems to hail from a distant planet—the traditional "other" of a different race or status, ethnicity or history—to someone just like you but who's seen things you haven't, illuminating things that alter one's path in the world. All of this, of course, is ancient advice; the shoes are a favored metaphor that underpins everything from "love thy neighbor as thyself" to "know thine enemy."

The key is picking some good shoes. And there are some, right now, in the summer of 2006, walking through America. One pair belongs to Rolf Mowatt-Larssen. He's the man who tries to figure out whether terrorists are about to wrap, or already have wrapped, their hands around this era's Promethean fire—uranium or plutonium—in sufficient quantity to make a nuclear weapon. He's been the government's leading man on this twisting, harrowing task since 9/11, first at CIA and just recently at the Department of Energy. That's means he's spent nearly five years regularly briefing the president and the vice president and kicking down doors here and abroad to ask unsettling questions. Either he's the most important man in the U.S. government or he's Chicken Little.

He's not sure himself, which is why his shoes are good for walking. You don't know where they'll lead. Neither does he. When he finds out, we'll all find out—and hopefully it'll be before there's a catastrophic event.

Right now his feet are up. It's late on Friday afternoon, July 14. He's reclining at his office in the basement of the Department of Energy, where he's the head of intelligence—a division that will grow from 100 to 350 by year's end—glancing over the dispatches of the president's meeting with Vladimir Putin. The Russian president, proudly hosting the G8 conference in Russia for the first time, wanted to start the gathering on a cooperative note. This was accomplished by signing an agreement with the United States to colead a multilateral effort to combat nuclear terrorism.

Rolf had reviewed drafts of the agreement, though he's not hopeful it will do very much. There are several agreements, much like the one Bush and Putin have just signed, that have been admirable in intent but severely lacking in follow-through and execution. They depend, as with all agreements, on the enthusiasm of the signatories. If Russia and the United States—the old cold war combatants who brought these weapons to the world—are not working together, not much will occur. And the relationship is strained. The United States says it wants to help Russia secure its stockpiles of material. The Russians say, with a note of impatience and resentment, that they can handle their own affairs.

Rolf, who did two tours for CIA in Moscow and helped catch moles like Aldrich Ames, doesn't draw much comfort from Russia's assurances. He knows too much. "Know thine enemy" has, of course, a modern, institutional translation. It's called "intelligence."

The most valuable intelligence has always been human intelligence—spies, or moles, in the opponent's camp—which has helped shape conflict since the Trojan War. Rolf happens to be one of the few people, anywhere, who has "run" operations against the great enemy of the past, the Soviet Union, and against this era's pernicious opponent, al Qaeda.

He thinks almost every day about the many lessons the United States learned in decades of recruiting and managing Russian spies, and if any of those hard-won lessons might be relevant right now.

The United States, mind you, has never been very good at espionage, and it still isn't. America has developed only three spies inside the al Qaeda terrorist network since 9/11. One, codenamed Lovebug—a Kurd and an associate of Iraq's al Qaeda chief, Abu Musab al-Zarqawi—signed on in 2002 and worked hard, offering contextualized intelligence worth terabytes of electronic surveillance. He left, vanished, and then returned to the fold just a few months before he was killed in early 2004. Rolf helped manage Lovebug, and he thinks often about why Lovebug came back. What was going through his head, and how do we find his successors? The world is too big, destructive materials are too widely available, and the footprint of a terror cell—maybe only a few guys in some apartment—is just too small. Without another Lovebug, several of them, telling us what's being planned, a nuclear event in a major U.S. city is not a matter of *if.* It's a matter of *when.*

If you want to walk in Rolf's shoes, you're going to hear a lot about walking in the shoes of *his* top-priority "other"—namely young Muslim men whose decisions, one by one, are driving global events. Sitting in his office, he tries to connect a few dots about what the terrorists may be planning. He flips through an array of classified dispatches of electronic intelligence. He looks over interrogation transcripts from Guantánamo Bay. He knows, more surely with each passing minute, that we are blind without moles, that al Qaeda is looking for uranium and there's plenty of it out there. With as little as thirty-five pounds, a sophisticated group could build a Hiroshima-size bomb. We'd never see it coming.

That's why he can't sleep.

He calls his wife. Tells her he'll be home late, asks what his teenage girls are up to this Saturday night, and just hearing her voice—a woman, God bless her, who's traveled at his side through five foreign CIA tours—reminds him of what he hopes.

He hopes he's Chicken Little.

Halfway around the world, George W. Bush walks to the end of the driveway of his villa and looks up at a blimp, full of electronic equipment, floating a few hundred feet overhead. It's Saturday, July 15. He stands there silently on the blacktop, waiting for Putin to pick him up in a golf cart and drive him to the day's events. Both of these things irritate him. Bush doesn't like waiting and he hates when other people insist on being in the driver's seat, which is where Putin has been throughout their relationship. He knows now that it might not have been that way. He was warned, and he didn't heed the warnings. In fact, they came months before his famous meeting in June 2001 where he said he felt that he was looking into Putin's "soul." He had talked to Putin several times before that, even during the presidential campaign, and he felt the man was his friend. Once Bush arrived in office, various Russian experts at the NSC tried to warn otherwise. They said that Putin was a trained KGB agent, and a good one. He wants you to think he's your friend, they said. That's his skill. Soon enough, they had a remedy. Putin was going to a meeting in Vienna in February 2001. He'd be staying in the presidential suite at the Hotel Imperial. CIA had an old listening device implanted in the wall of the suite. All they needed to do was replace the battery.

Bush is a guy who needs to make things *personal*—it's how he's always organized a complex world—and he felt that he'd developed a bond with Putin. When CIA made its offer, his response was that you don't wiretap a friend. Condoleezza Rice said it was "too risky, it might be discovered." CIA said that if it was, it would probably heighten Putin's respect for Bush. Bush settled it—it was a gut decision. No dice.

This was an early sign of an extraordinary dilemma, one that would come to define America's posture in the world: Bush's powerful confidence in his instinct. It might be called a compensatory strength, making up for other areas of deficit. He's not particularly reflective, doesn't think in large strategic terms, and he's never had much taste for the basic analytical rigors embraced by the modern professional class. What he does is size up people, swiftly—he trusts his eyes, his ears, his touch—and acts. While he has an affinity for stepping inside the shoes of others, his métier is often brutally transactional rather than investigatory or empathetic: he is looking for ways to get someone to do what he wants, and quickly. This headlong, impatient energy fueled his rise, as anyone knows after watching him strong-arm a big-money contributor, parse friend from foe, or toss a script and preach, heart to heart, to supporters in the Republican base. It's how Bush—like many bullies who've risen to great heights—became the president.

Once he landed in the Oval Office, however, he discovered that every relationship is altered, corrupted by the gravitational incongruities between the leader of the free world and everyone else. Everything you touch is velvety, deferential, and flattering. To fight this, presidents have been known to search furiously for the real, for the unfiltered, secretly eavesdropping on focus group sessions far from Washington, arranging Oval Office arguments between top aides—a Gerald Ford trick—or ordering policy advisers, as Nixon often did, to tell them something the advisers were sure they didn't want to hear. These men, even with their overweening confidence, embraced a unique kind of humility, recognizing they were in a bubble and fearing they would make historic mistakes.

Bush, with his distaste for analysis and those who contradict him, didn't go down those paths, and he seemed unconcerned, unlike other presidents, that isolation would prompt errors in judgment. Instead, he began taking policy advice from old cajoling friends whose relationships predated his ascendancy or from visiting pastors speaking frankly in their *everyman* voices of faith. A man who trusts only what he can touch placed in a realm where nothing he touches is authentic. It's a diabolical twist worthy of Sophocles or Shakespeare. Either would have written it as a tragedy.

Because, over the years, the bullying presence of

Bush—making things *personal* without hesitation or limits—became the face of America. It was an effect caused by much more than his confrontational public pronouncements. Despite his advisers' admonitions that his relationships with other world leaders were, like so much else around him, manufactured, he felt that they were real and easily managed; that foreign leaders would submit to his persuasive charms and persistence and do what he wanted, even if it was against the interests of their countries. Blair was "a good man, a friend, who got it"; Saddam was "the guy who'd tried to kill my Dad"; as for Putin, "I was able to get a sense of his soul," Bush said. He's "straightforward" and "trustworthy."

When each man defied the will of the president—in small ways or large—Bush saw it as disloyalty, and responded in unproductive, gut-driven ways. After eventful years and Bush's reelection, the nation and its leader became inseparable, as America, itself, was viewed as angry, reckless, petulant and insecure, spoiled and careless, with a false smile that concealed boiling hostility.

All this leaves Bush—and America—with limited options by the summer of 2006.

Yesterday Bush and Putin launched the Global Initiative to Combat Nuclear Terrorism, a small multilateral step in the right direction that seems to augur well for future collaboration. But in other venues, the Russian president's been busy telling foreign leaders to beware of the American hegemon, a fervent complaint, a challenge, that is drawing an enthusiastic response from countries who feel that they've been bullied and from the pride-starved Russian people. Meanwhile, various uranium-smuggling networks are operating in Russia, and al Qaeda, ever ready to buy, is rapidly reconstituting itself in the lawless tribal regions on the border between Pakistan and Afghanistan. If Bush had seized upon that early wiretap opportunity, he would have heard Putin's real voice and been confronted, from the start, with the hard reality that presidential relations were not personal. They were strategic, with Putin and other world leaders, state-based or otherwise. While Bush has spent years relentlessly questioning intelligence briefers about the personality of bin Laden and Zawahiri, wondering, "What makes them tick? What's driving them?"—trying to make it personal so he can engage forcefully, emotionally—he won't acknowledge, even at this late date, that these men he hates, these religious fanatics, are executing a fairly elegant strategy that can be countered only with care and dispassion.

And with all the setbacks, now well into his second term, the president still won't admit—even to himself, it seems—that you can't run the world on instinct from inside a bubble.

The president looks at his watch and then up at the blimp. Yes, he's been waiting like some kind of idiot in the driveway for five minutes. A reporter from *Newsweek* approaches gingerly and asks if Putin wears his "dour KGB face" in their private meetings or if he's more relaxed. Bush points up to the spy balloon and quips, "That's your phrase, not mine." Just then, the Russian screeches up in his golf cart and they're off, with Rice in the backseat, for a day of meetings and press conferences, where, hour after hour, Putin displays his assiduously earned leverage.

That night, George and Laura Bush, along with seven other world leaders and their spouses, settle along a vast, spectacularly adorned table inside the Peterhof Palace, built by Peter the Great, as waiters in powdered wigs serve a seven-course meal of lobster and beluga caviar and beef Stroganoff with truffle sauce. Putin looks admiringly across the table. Everything is carefully crafted. All relationships at this level, after all, are strategic and manufactured. This gathering is a moment of triumph for him.

Outside the palace, a trained bear in a pink polkadot tutu had regaled the guests by walking on its hind legs and performing somersaults, and now Germany's chancellor, Angela Merkel, offers an appropriate story, clearly tested by her aides in advance, about the recent shooting of a rare wild bear in Germany.

This prompts a surprising response from Japan's prime minister, Junichiro Koizumi. In a flight of free association, he begins to search aloud for every English word or phrase he knows that contains the word *bear*. He must be improvising. How rich.

"Teddy bear," he says, serenely, as world leaders begin to chuckle.

"We must *bear* criticism."

More gaiety. Yes, go on!

He pauses.

"Un*bear*able." And the room dissolves in laughter.

Back in America, in Northern Virginia, a longtime U.S. intelligence official is puttering through his Saturday afternoon. He's running errands for his wife, something bosses in the intelligence community haven't done very much of since 9/11.

But there's less and less to do for those who've worked years in America's clandestine service recruiting sources and running undercover operations. CIA, as America's primary intelligence agency, doesn't exist as it once did. It now lies beneath the Office of the Director of National Intelligence, a fast-growing, thousand-plus-employee agency, and is ever more insubstantial beside the growing intelligence division of the Department of Defense, which controls 80 percent of America's $50 billion annual budget for intelligence. Old agents and intelligence managers in both halves of CIA—operations and analysis—have fled to private contracting firms. New recruits have been hired in a hurry. Half the agency's workforce now has five years' experience or less.

The official's been there for decades and was there, on site, during America's brief springtime of robust human intelligence—the year or so after 9/11—which ended with the military campaign in Iraq, as anti-American sentiment became the currency of global opinion and terrorist recruitment skyrocketed. The United States hasn't caught a top terrorist of any real value in two years. Even if many Muslims hate al Qaeda, they don't want to help the United States. If someone were to see something pertinent in Pakistan or Saudi Arabia or Yemen these days, he'd most likely look the other way. Let America get its comeuppance.

This state of affairs is untenable—a loss of intelligence capability that will end in disaster—and he thinks about why, about how America went from a country that people wanted to help in its time of need to one they'd just as soon see humbled.

And each time he goes through this exercise, he comes back to Iraq and the suspicions of so many, in the United States and abroad, that we went to war under false pretenses. He thinks it's the key reason the United States has lost its moral authority in the world. People—at the agency and around Washington—dismiss it, the whole mess, saying it's all past tense, let it go. But he knows more than they do—more than all but a dozen people, maybe fewer, inside the U.S. government, with two of them being Bush and Cheney.

He knows there was a secret mission a few months before the war—a top-drawer intelligence-gathering mission that the United States was involved in—that found out everything we later learned. That there were no weapons. And we knew in plenty of time.

But what he knows—this troubled public servant—is itself only a glimpse of something much larger, and still submerged.

It is a violation of American principle and law that lies, quiet and sure, beneath the country's misfortunes. And, like a demon, it must be exorcised before dawn.

The next morning, a sunny, cloudless Sunday, Candace Gorman's low heels crunch across the sandy paths of Guantánamo Bay.

She tries to make conversation with the guard escorting her; he'll have none of it. He's a young soldier, tall and blond, and he seems angry at her. He leads her silently to a small second-floor room in the complex and locks her inside.

The man sitting at a table across the room, his leg chained to the floor, looks quizzically at her. "How do I know you are who they say you are?" he says in serviceable, accented English. "Maybe you're someone here to trick me."

Candace fumbles through her purse and hands him a business card. He shrugs. "Anyone could have printed one of these."

All attorneys registered in Illinois have to get a new bar membership card each year, and Candace, a pack rat, has kept them all. She digs through her briefcase for her official ID. She finds it after a minute, and last year's, too. And the one from the year before. Five minutes later, twenty-six laminated cards are lying on the table. She's had the briefcase for a quarter century.

"Welcome, Mrs. Gorman, and thank you for coming here. I have imagined it."

"Thank you, Mr. al-Ghizzawi, I am *officially* your lawyer." With that exchange of consent, Candace Gorman, a fiftyish civil rights lawyer from Chicago, mom of three teenagers, steps to the edge of a border, a low, long table separating her from a man the U.S. government calls among the "worst of the worst."

They settle into this odd, longish room, used for

both interrogations and attorney visits, with its small table, two chairs, and cell in the corner—an eight-foot-square cage with a cot, a toilet, and a door, now open. Candace pulls out her file, which contains two notes her client has sent her about his declining health. "How are you feeling? That's the first thing."

Ghizzawi sighs, and begins to list his ailments and their history. His health began to decline in 2004, and he's been in increasing pain ever since. He's been vomiting constantly; his stomach is raw. He's lost about forty pounds. There is pain in his left side, in his back and his right leg—the one chained to the floor beneath his chair. Candace watches him carefully as he speaks. He's about five feet ten—but he can't weigh more than 120 pounds. He's pale, yellowish, and weak.

She doesn't want to ask him too many questions to start, thinking about how many have already been thrown at him in interrogations. So she talks about herself and mentions that she works on cases involving civil rights. The term seems unfamiliar to him.

"In the U.S., we have rights that people have to be treated the same regardless of their religion, their race, or whether they are a man or a woman, and these rights are defended by laws. If a company or the government breaks those laws in regard to someone, I represent them. I file lawsuits in the federal court."

He nods, tentatively.

"And I think your rights are being denied, because you have the right to at least know why you're being held." But before that sentence, about rights denied, is halfway spoken, Candace's mind seems to slip backward, locking onto something she'd buried during the months of principled debate and legal struggle just to get here: this man might actually be a terrorist. Her victims of race, age, or sex discrimination were just working people. Mr. al-Ghizzawi could be Taliban, or even al Qaeda. She pushes the thought from her mind. Beside the point. This is about due process, about letting the law do its work.

First, she needs to get his story straight, and she does: How he was a baker in Afghanistan who moved with his family, as the bombs fell, to a new town. As a Libyan and a stranger, he was summarily accused of being a terrorist and handed over to U.S. officials in early 2002 in return for a sizable bounty.

Candace fills one legal pad after another. Ghizzawi becomes more engaged, hour by hour, wanting to know everything about Candace and her family. She says her father, nearly ninety years old, has recently become ill, and they talk about that. "You're lucky to have a father who has lived such a long, full life," he says. She smiles—yes, that's true—and then tries to keep that slippery thought (*He could be a terrorist*) in her grasp.

"One morning," he tells her on the second day they're together, "I saw a tiny flower, a rose, growing in the sand just beyond the bars of my cell. I thought, I am like that rose. Neither of us belongs here."

Candace feels her eyes welling up. *Shake it off.* "So what are you reading?" she says, changing subjects. There's a book lying open on his cot. He complained earlier about not being able to get the reading materials he wants, especially scientific or medical books to assess and perhaps diagnose his medical condition.

He tells her it's *Moby-Dick*, an abridged version in both Arabic and English.

She's delighted—tells him, "It's one of the most famous books in the English language"—and they talk about the plot and characters.

"I know it is a very good book, very famous," he says. "But I don't understand, why does Ahab want to kill this whale so very much? It was just a leg. He only lost a leg. It's not like he lost his arms or his family was killed. He is still able to be a captain. Why so much vengeance to get the whale?"

"Well, he's obsessed."

Ghizzawi shrugs. "This guy just doesn't give up. I don't understand."

She looks at him, befuddled. Soon she'll be leaving; they've been talking for nearly three days. And nothing is adding up. Could he be a terrorist and not understand Ahab's obsession with his target? Or a hard-bitten Taliban fighter who finds common cause with a rose? She snaps to and bites down hard on reasonable suspicion.

But it's all moot. Guilty or innocent, this man deserves the due process of law. Period. She tells him, matter-of-factly, that she plans on filing a variety of appeals to get at the evidence arrayed against him. Might take a month, maybe two.

He watches her talk, his feet, filthy and blistered, shuffling in green foam flip-flops, like the ones kids wear at the beach. From the talk of plans and dates, he knows she's leaving. "I really need medical help," he says, softly. "This is what I need the most. I'm afraid. Please, I don't want to die here."

The world stops, for an instant, as a man, a Muslim who's been swept around the world by history's currents to a jail cell in Cuba, pleads for mercy to a woman from Chicago, a lawyer, a mom, and a Christian, who, suddenly off balance, tries to grab hold of principles formed across millennia—through Hammurabi and Blackstone and Learned Hand—to break her fall. It is faith in the law she turns to. It's all she can offer really, she thinks, as that Aristotle quote—a law school standard about how "the law is reason, free from passion"—pops into her head.

This is *not* personal.

But, of course, it is—as is this intensely personal book. It's about how people, in America and abroad, are trying to grab hold of what may be one of the most powerful forces on earth—moral energy—which flows, most often and most fully, from the varied and connected chambers of the human heart.

Dig deeply enough and it becomes clear that the great public institutions, and the law itself, are actually built on the most intimate of human qualities, such as honesty or forgiveness. Each of the characters mentioned thus far is driven—sometimes unwittingly—by very basic human values. Rolf is stumbling forward, sleepless, looking for simple truths to help him burn off fear. Candace is trying to find a place within the law for a powerless Muslim man and to prove, at least to herself, that justice is maybe more about compassion than judgment. George Bush has an important role, too. As people lament America's diminished moral authority and point to the president's actions and image as the cause, they are using Bush as a fixed point—something to push against—in charting new paths. That would apply, for instance, to the intelligence official who carries an enormous lie in his churning gut. It is one of the great lies in modern American political history. He wants simply to say we're sorry, and we'll learn from our mistakes, as all truly great nations must.

MORSE INSTITUTE LIBRARY
14 East Central Street
Natick, MA 01760

It's a sad turn at a time of crisis—as religious fanatics search tirelessly for history-ending weapons—that the conduct of nations and of leaders, both duly elected and self-appointed, is so unimaginative and self-defensive. And that America has lost so much of the moral power that the world now desperately needs it to possess.

That dispiriting truth is something the characters in this book, like so many others across the planet, have begun, at long last, to see as a starting point.

It's a starting point for their journeys, each quite personal, to find a moral compass, and a way forward.

MORSE INSTITUTE LIBRARY
14 East Central Street
Natick, MA 01760

ACT I

Other People's Shoes

Chapter 1
Welcome Home

Usman Khosa awakes to the voices of his room-mates in the kitchen. A hazy sun is shining in, giving the exposed brick above his bed an orange hue. He checks his night-table clock—7:15—and slips back into the deep sleep of a young man.

It is morning in America. Or at least in an apartment near Dupont Circle in Washington, D.C., where three young, well-educated men start a summer's day. They are friends, a few years out of Connecticut College, dancing through the anxious glories of first jobs and few obligations. It's a guy's world. Linas, a strapping Catholic, American-born, with Midwestern roots, is an economic analyst; David, Jewish and gay, with wavy brown hair and movie-idol looks, is a public relations staffer for an international aid organization. After breakfast, they slip out together, each in a blazer and khakis, Christian and Jew, straight and gay, into the flow of the capital's professional class.

Their Muslim roommate hears the front door shut and rises with a sense of well-being. He'd worked late, as usual, and then met some friends for dinner, a night that went late with loud talk and drink. He came south to D.C. from Connecticut just over three years ago, a day after receiving his diploma with its summa cum laude seal, to a waiting desk at an international economic consulting firm, Barnes Richardson, with offices across the street from the U.S. Treasury Department and a block from the White House. He finds the work fascinating because it is: taking sides in bloodless struggles between countries and their major corporations over product dumping and tariffs. Trade wars. It's the kind of conflict that smart folks thought the world was moving toward in the mercantile 1990s, when the Soviet Union's fall was to usher in a post-ideological age, a period when aggression would be expressed, say, with tariffs on imported cars and wheat dumping. It was a hopeful notion that issues of progress and grievance, the fortunes of haves and have-nots, would be fought on an economic field where the score could be kept in terms of GDP, per capita income, and infant mortality rates. It wouldn't turn out that way, as the few who saw the rise of religious extremism foretold.

And that's why the boy brushing his teeth this particular morning—July 27, 2006—is not just any young professional on the make. He is, notably, a Muslim from the fault line country of Pakistan—the home, at present, of Osama bin Laden and Ayman al-Zawahiri, Pervez Musharraf and Mullah Omar, fifty-five nuclear weapons and countless angry bands of Islamic radicals. Usman, from this place, of this place, strives with an ardent, white-hot yearning to be accepted into America's current firmament of fading hopes. Like each fresh wave of newcomers, he presses mightily to make that hope new. Whether he means to or not, he's testing American ideals at a time of peril.

It's a fault of cultural nearsightedness, or worse, that he is not immediately seen as identical to immigrants glorified in oft-told tales of potato famines or Russian pogroms or, back further, a search to worship freely by some *Mayflower* stowaway. He is, after all, identical to them in every essential way.

But his journey involves a blue '78 Toyota Corolla. In Pakistan, a car is a symbol of a man who can move as he wishes, where he wishes. A new one is a rarity, a luxury, and Usman's father, Tariq, was given the car as a wedding gift from his father, who told Tariq that a married man should have a car, and he should be his "own man, beholden to no one."

The Khosas have a deep history in the region that now lies at the geographical heart of modern-day Pakistan, but the family is not among the few dozen elite who long ruled South Asia and cut deals with the British when the empire took over in the 1860s. The shaping hand of the Brits is still keenly felt in the region, particularly in its cutthroat academic tradition. *Competition* would be too generous a word. It was more emancipation through recitation, a test of classical British learning with a million contestants, a handful of winners, and enormous prizes, all determined by a crucible known as the civil service exam. In the vast country of India, a fraction of the highest scorers would win coveted acceptance into the civil service—the bureaucracy, running their country for the British—which came with grants of significant leverage over their countrymen and subtly stolen rewards. Even after India broke free in 1947, the civil service test remained, grandfathered in by the country's ruling elite, who could recall the posting of scores—the day, the minute, the sensation—like a family's second birth, cited often and judiciously from parent to child across eras.

Usman's grandfather, a very good student, finished one slot out of the money, so to speak, but carried the fervor of the runner-up into the newly created state of Pakistan. As a young man, he met Muhammad Ali Jinnah and thoroughly internalized the great man's vision of a Muslim state that would break away from Hindu-dominated India; an Islamic republic with mosque-state separations and protections modeled loosely on Western democracies, where religion would be largely a private matter and rigorous education all but deified. Jinnah's idea was that this balance would allow the growth of a professional class that would become the country's cornerstone of progress. Usman's grandfather embodied that vision. He became a lawyer, involved himself in countless public causes, and began to sell what land the family had built up in the past few centuries to educate his children in the finest regard Pakistan had to offer. Usman's father, Tariq, was the eldest and the first beneficiary, taking his college degree and that blue Toyota on an array of edgy professional missions and rising through Pakistan's competitive bureaucracy to become one of the leading law enforcement officials in the country. Like many bureaucrats, he moved between government houses, even had government servants, but acquired little cash, and so the remainder of the family's land was sold to educate *his* children at Pakistan's best schools. This meant that Usman's sister, two years his senior, starred at Lahore's finest private academy for girls and won a full scholarship to the London School of Economics. And that Usman, a blazing student at Lahore's exclusive Aitchison School—built a century before by the British to educate the children of India's feudal families—was given a full scholarship to Connecticut College. The problem came down to what wasn't covered: the costly flight from Pakistan to America.

After twenty-two years of faithful service, the Toyota spoke to Tariq. He'd invested an abundance of attitude and nostalgia in the old blue beast, buffed it regularly, scraped out rust; he could feel the distance traveled, for both car and driver, in the sag of the chassis, the glossy bareness of the upholstery. Everyone knew what the car meant to him, and what it meant when he sold it for his only son's plane fare to America.

That's how the Khosa line—Jinnah's line, in a way—passed to Connecticut, where Usman studied fiercely, headed the Muslim Students Association, and became a leader in the student government. He met his current roommates as a sophomore presiding over the freshman class's disputed student government election, in which both Linas and David were candidates. They both ended up winning their races, and all three now see this as rich and ironic, that Usman—hailing from the due process–challenged Pakistan—was the Connecticut College election commissioner who handed out victories. They furthermore think it's "sitcom-worthy" that a Jew, a Christian, and a Muslim are sharing an apartment as the world's us-versus-them divisions seem to be boiling over. They argue, with fierce good nature, over who should do the dishes or whether Usman should introduce Linas to a nice Muslim girl or find a nice Muslim boy for David. No loss of confidence in the cross-border ideal, not here. In fact, this three-bedroom apartment—galley kitchen, utilities included—is a safe house of sorts, the opposite number to a cell of young religious radicals arguing over the dishes in Wembly or Karachi or Kabul.

Usman, like immigrants before him, is a walker. It's something about the crowd, its intimacy and anonymity, and the way you can flow inside of it. It draws him in. Any trip of a few miles or less he takes on foot. He keeps business suits and sport jackets in the closet at work, or at the dry cleaners near his office. So each day of summer he slips into shorts, Nikes, and a T-shirt and squeezes his laptop into a backpack.

It is a warm day, but not nearly as warm as Lahore, he thinks, stepping outside. Oh yes, Lahore is much hotter than this, and dusty. Washington, even at its most humid, feels temperate and superior, lovely and fresh, and seems to wash ancient grit from his pores. He stops on the top stoop, spins his iPod to a play loop of Arabic tunes, and sets forth for Pennsylvania Avenue.

"Thank you, that's fine for today," George W. Bush says, as he dismisses a half-dozen attendees of his morning intelligence briefing and settles behind the world's most famous desk. He's agitated, doing his best to get things in order before he leaves for his annual August vacation in a week. But the world won't heed his will, not anymore. The Oval Office is quiet—an unscheduled half hour—and a precious moment to step back, to take stock. His best-laid plans for this summer are already in tatters. It was to be a season to focus on his strengths, with the midterm elections just over three months away. That meant domestic issues, where he has capital with a reasonably strong economy, and events highlighting his one remaining area of strength in the foreign arena: handling terrorists.

Except everything, and everyone, has been conspiring against him. His poll numbers are in the basement, with several mid-July tallies putting his approval rating at just 40 percent, the lowest for any modern president going into the midterms. Casualties in Iraq have been steadily rising since the spring—the country is all but exploding in sectarian violence. Karl Rove and Condi Rice are talking about shifting the rhetoric on Iraq away from the value of America's eventual triumph to the unthinkable dangers that would attend America's withdrawal. He spent yesterday, July 26, with Iraqi prime minister Nouri al-Malaki, who gave a speech to Congress that the White House staff worked and re-worked until it screamed. The consensus sentiment in the morning's papers is that al-Malaki gave a campaign speech that was completely divorced from reality.

If people want depressing reality, there's plenty of that to go around. Israel is sinking deeper each day into a disastrous engagement, now two weeks along, with a stronger-than-expected Hezbollah. It's a mess. The morning's reports from the region show the worst day of Israeli losses yet—9 dead—and ever worsening PR blowback from two weeks of "unintended" casualties, now at 489 Lebanese civilians. Meanwhile, only 20 Hezbollah fighters have perished. And, two days ago, 4 United Nations workers died when a clearly marked UN outpost was hit by Israeli bombers. Reports have emerged of the humanitarian workers madly radioing Maydays to the Israeli army in their last moments.

Bush talked this morning at 7:30 to Angela Merkel, the German chancellor, about all this. The unified front on Israel and Hezbollah they both helped craft at the G8 meeting last week in Russia is in tatters. Yester-day Rice was in Rome, where representatives from the United States, Europe, and the Middle East met to try to hammer out a cease-fire. But the terms were untenable, and Bush, talking to Rice, opposed it. He was instantly pilloried for that in last night's news cycles, and all the G8 partners—everyone except the Brits—are distancing themselves. So when he talked to Merkel, he told her no one is saying that aggression is the first choice, not for the Americans in Iraq or the Israelis in Lebanon. But it must be an option. In his first National Security Council meeting as president he tried to set the tone, telling all the NSC principals his view that "sometimes a show of force by one side can really clarify things." Couldn't have been much clearer than that. Said it again this morning to Merkel, and still believed it, more strongly than ever. Merkel at least understood this position. She wasn't like her predecessor, Gerhardt Schroeder, who opposed the United States on Iraq and didn't seem to think force was ever justified. No, Merkel understood. She'd said so on the phone—their second call this week—and that she would make statements in the coming days in support of the United States and Israel. Some things were worth fighting for.

But what's really driving Bush's calculations at the moment is the just-finished intelligence briefing. It involved a plot that he's been hearing about for some time. The British have been working it since last year: a major terror cell in the suburbs of London. While it's their case—they've made that very clear—U.S. involvement has deepened as the tentacles of the cell have spread across Britain to Pakistan. With about forty suspects, sending plenty of e-mails and making calls, the Brits have increasingly had to rely on what Bush likes to call the "firepower of Ft. Meade," the massive National Security Agency surveillance complex in the Maryland hills.

Then, in the past few days, everything changed. Electronic surveillance revealed, finally, the nature of the plot: airliners taking off from Heathrow carrying explosives and headed for the U.S. East Coast. Talk among the suspects revealed it could involve as many as a dozen planes blowing up over U.S. cities. That would make it the biggest plot since 9/11—the so-called second wave that Bush and Cheney have been waiting for all these years. Reports of all kinds have been coming to Bush's desk as the U.S. anti-terror machine has secretly ratcheted up. Mike Chertoff, the head of the Department of Homeland Security, is working his intelligence unit around the clock; NSA's working overtime; and CIA is doing what it can, especially with its sources in Pakistan.

This morning's briefers said that the British are advising the United States to sit back and take a deep breath. The Brits have been stressing that this might be just early logistical talk, that they have these suspects so completely wired that they can't sneeze without generating an electronic dispatch, and that no one is doing anything that would pass for an active operation. Bush has heard this before. Patience, *patience*. The British are saying that all the time; and that they're better at intelligence work than the United States—they've been doing it longer, they have experience with the IRA's terror network, and they're especially well placed in target communities, such as the Pakistanis and the Saudis. The United States, with all its electronic firepower, is having more and more trouble in recent years with the basic spy craft of recruiting spies and getting actionable information from walk-in informants. The big breaks, of course, have come from sources on the inside or nearby, sources that took time to develop, and from informants in communities close to the action. The United States is too anxious and trigger-happy, the Brits complain, taken to picking up some bit of an overheard conversation and then sweeping up suspects. Blair said in a recent conversation with Bush that this was "the error of relying on the capability you have rather than developing the capability you need."

The Brits, after their experience in Northern Ireland, were starting to believe that the key was to treat this not as a titanic ideological struggle, but rather as a law enforcement issue. This required being patient enough to get the actual evidence—usually once a plot had matured—with which to build a viable case in open court.

But waiting didn't feel right to Bush, not now. It could take six months or more—who knew?—until this plot became operational. Blair was flying to Washington late tonight. They had a full morning planned for tomorrow, a long meeting and then a joint press conference. Blair was a good man, and they'd covered for each other plenty of times. Blair would come through.

Bush looks up at the Tiffany grandfather clock in the far corner. Ten past nine. He has to leave for a bill-signing ceremony on the South Lawn in five minutes, then he meets the Romanian prime minister, in town for a visit, then they'll do a press conference together, then lunch, and then he has an economic speech to give in front of the National Association of Manufacturers. He flips absently through a briefing book with some talking points for the day and forces himself to focus. For him, it's always been a struggle between the analytical and the emotive—the former, an effort; the latter, so natural, so clarifying. His feelings, his hunches, have gotten him and his nation into some tight spots. He's aware of that. So he's tried to be more attentive lately, tried to read the briefing books—to study them, with their seasoned, prudent, boring-as-hell advice.

What no one understands, no one but Cheney, is how hard some days are. People are not bending to his rightful desires as they used to. He remembers what it felt like, in the two or three years after 9/11, to possess native authority, and he misses it.

But in this one area—the secret world of intelligence, of foiling terrorists and their plots—he still has control. Everything unfolds in shadows. The results are what *he says they are*. Now that they have the terrorists on tape saying their plot is directed at America, he wants it shut down, ASAP. This is what he, and his Republican Party, gets paid for—protecting America—something the voters might benefit from being reminded of. What could be more important?

Through the leaded glass windows of the Oval Office, a limousine and four black SUVs idle in the driveway near the portico—Cheney is gearing up to deliver a speech at the Korean War memorial on the Mall. The July sun is burning off a morning mist. It's going to be a hot one. Bush puts on his suit jacket, ready for his day, as the motorcade roars toward the gates.

Usman Khosa is walking east along the White House's ornate wrought-iron fence, tapping it with his hand, like one might tap a white picket fence. A U.S. Park policeman moves toward him. The officer is saying something. Usman can see his lips move, shouting. He pulls out his earplugs. Arab music blares. "Cars will be passing out of this gate. You need to stop!" Usman nods, apologizes, and fusses with his iPod as other walkers back up behind him. The gates slowly begin to open. He looks up at the majestic building—he loves looking at it. He takes a route each morning down Sixteenth Street, where the White House is visible from a mile away and grows larger with each passing block. It makes him feel consequential to walk toward it, like he's going to meet Bush, or could, and often thinks about what that'd be like. Bush is the most known person in the world—his presence, his face, has become the face of America; his pointing finger, punching the air, an emblem of the way the United States engages with the world. Usman feels he knows the man and that, as a Muslim and a lover of America and someone who often finds himself reluctantly defending Bush to invective-spewing Pakistanis, he and Bush might actually have a worthwhile conversation.

He's more right about this than he realizes. Though Bush represents America electorally, his life is as *un*-representative as it could be, a cocoon of rare luxuries and sycophants, a carefully crafted, busy, filtered array of activity, much like the lives of those sitting atop so many venerable institutional mountains, only more so.

But those snowcapped peaks are, in the modern age, melting—the range itself, crumbling—under pressure from the shifting, heated, quaking landscape of the individual. The information age, after all, is the age of the individual, a time when great waves of personal choice and expression—magnetic fields of impulse, connected on a borderless, global grid—can level the assembled power of armies and nations. The now common term *asymmetric* suggests that the comfortable symmetries, strategies, and conversations between those on the mountaintops are increasingly inconsequential as power shifts from the peak to the base, not only to petulant, entrepreneurial "rogue states" but also all the way to the bottom, to those—many among the multitude— who are *representative* in their experiences, their sensations and swift judgments. They move in vast heartbeat migrations, both sensing nascent currents—anger, fear, hope, rage—and creating them in a kind of dizzying simultaneity, open-source and dynamic, that overwhelms institutional norms of fact gathering and review and analysis. It is on this landscape—no less startling for now being familiar—that a president laments that he gets better, more immediate intelligence from CNN than CIA; where some religious purist who posts a beheading video—downloaded endlessly, driving news cycles—is the black-sheep cousin to a college-dropout-cum-billionaire who creates Facebook.com; where two guys in a cave confound the message and might of the world's most powerful nation; and a few ardent twenty-somethings, with modest, acquirable skills and stolen credit cards, could assemble a device that wipes out Midtown Manhattan.

Usman Khosa, as a young Muslim man in search of a better life, busily ingesting the widely available fare of modernity's mishmash, is *representative* in ways that are particularly consequential at this moment in history. Like it or not, he—and countless others like him—is in a discussion with the isolated man memorizing "talking points" in an oval office 164 yards away.

And, in present tense, the ornate gates are now open. The limo and security SUVs speed out of the delicately cobblestoned roadway between the White House and the U.S. Treasury Department. Usman and a dozen people who've gathered in the past minute watch them pass and attempt a futile glimpse though tinted glass. Usman fiddles with his iPod as the gates close and begins to walk among the crowd toward Treasury.

Right in front of the statue of Alexander Hamilton, at the base of a long sweep of steps leading to Treasury's neoclassical pillars, he thinks he sees a flash of white out of his left eye. A bicycle is being flung. He turns as a large uniformed man lunges at him.

"The backpack!" the man yells, pushing Usman against the Italianate gates in front of Treasury and ripping off his backpack. Another officer on a bicycle arrives from somewhere and tears the backpack open, dumping its contents on the sidewalk.

Usman is in a daze, spread-eagle, grabbing cool iron. "What? What!"

"Don't move!"

Pedestrians start madly dispersing, sprinting in crouches, hands over ears, running for cover.

Usman is mortified, breathless. He can't speak for a minute, maybe longer, as his bag is repeatedly searched and he's roughly patted down once, then again.

"You've made a mistake," he croaks.

The Secret Service officers look at him, unflinching. One talks into his walkie-talkie, calling for backup.

"Are you a U.S. citizen?"

"Do you have any weapons on you?"

"No, I'm from Pakistan. I have a visa."

"Do you have your visa on your person?"

"No, um, I don't carry it with me."

"It's illegal not to have it on you."

"I thought it wasn't a good idea."

"Do you have your visa control number?"

"No, I'm sorry."

"Do you have any guns or weapons on you?"

"No."

"Do you like guns?"

"What? No, not really."

And around they go. Minutes pass.

Usman sees a reporter approach, a woman, late thirties. She's gazing warily at him as she motions to one of the officers to huddle, and they cozy up. She says she's with the Washington-something—Usman can't quite make it out, the *Post*, or the *Washington Times*—and he sees, with crushing clarity, what she and the gawking crowd, now rimming an estimated blast radius the size of a baseball diamond, all see: pakistani terror suspect arrested at white house. The officer and she go around the corner to talk, privately, leaving Usman and the other officer alone at home plate. He wants to scream to her, to everyone, that he got an A in freshman English, that he's read *The Federalist Papers* from beginning to end—in Urdu, for God's sake—and can quote passages verbatim. Hamilton is standing right next to him. Ask Hamilton!

He takes a deep breath and tries to reason with the remaining officer. "My passport is right in this building, right here." He points to the office building over the officer's shoulder, just across the street. "I work right there."

The officer turns, a muscular uniformed man looking up at this smooth temple of white-collar privilege, and then turns back.

"You go to college in the U.S?"

"Yes sir, I went to Connecticut College, in Connecticut, from 2000 to 2004." Thank God, the miracle of education. *Keep talking.* "And, um, last year I did a semester at Dartmouth, Dartmouth College, at the Tuck School, a pre–business school program."

"Dartmouth?" the officer says.

Usman nods. A golden name, *Dartmouth.*

"So, what, you're some kind of smart ass?" The officer says this with a mocking tone, but Usman can't find the edge, the pertinent context. Is it a class thing, or racial, or something someone from Dartmouth once did to this guy?

"No sir, no, I'm not so smart."

Just over on the South Lawn, George W. Bush steps up to the microphone. "Thank you. Good morning. Welcome. Thanks for being here on this special day. Please be seated. America began with a Declaration that all men are created equal . . ."

Black faces glower from the mostly hostile crowd of five hundred gathered on the freshly mowed grass. After Bush gives this short speech, he'll sign the reauthorization of the 1965 Voting Rights Act, which finally put the full might of federal law and the federal courts behind the right of African Americans to vote.

If necessity is the mother of invention, desperation is the stepfather. The Republican Congress, seeing a downturn in their poll numbers, actually pushed through the reauthorization of the Voting Rights Act a year early to have something, anything, to court the black vote in November. The modern Republican Party, the party of Lincoln, has not been a favorite with blacks for decades—about 90 percent have voted Democratic since the New Deal. But the past few years have been notably miserable. Their current Republican standard-bearer, George W. Bush, got 8 percent of the black vote in 2000 and 12 percent in 2004, but after the Hurricane Katrina debacle in the fall of 2005, his approval rating among African Americans fell to an astonishing 2 percent. Early this year, Republicans in Congress appealed to Karl Rove; some had significant black populations in their districts, and Bush, along with the entire Republican leadership, needed to show a little effort. So the Voting Rights Act was reauthorized. And three weeks ago, Bush finally, after six years of demurrals, deigned to speak at the annual convention of the NAACP.

He hated doing it. He hates speaking in front of hostile crowds. Always has. It's the residue of being doubted since he was a boy—"the problem with Junior"—and on through the first twenty years of his adult life, as he limped through college and moved from one middling business to the next, until companies started wanting him because of his famous father. Becoming president himself, and then being reelected, was a kind of cathartic revenge, a shattering of all that. Then the doubts started to reassemble, surprisingly, shard by shard—particularly in the past two years with Iraq and Katrina. Doubters, now, everywhere he turned. The White House political team has worked harder than ever to keep him in front of friendly crowds—military, conservative, and, especially, religious.

Today is the exception, but they've conjured up a twist: cover those black faces with church fans, old-style, handheld Southern Baptist church fans, with voting rights act printed on the back. Considering it's eighty-one degrees by nine-thirty in the morning, everyone starts fanning, giving the South Lawn the feel of a big-tent revival, where someone might want to preach.

The religious trappings offer comfort, but no one is expecting much. Just adequacy, just get through it, and Bush continues his short speech on the terra firma of the Declaration of Independence, saying, "it marked a tremendous advance in the story of freedom, yet it also contained a contradiction: Some of the same men who signed their names to this self-evident truth owned other men as property. By reauthorizing this act, Congress has reaffirmed its belief that all men are created equal; its belief that the new founding started by the signing of the bill by President Johnson is worthy of our great nation to continue."

He botches that last line, but people applaud anyway—how could they not?—and it's enough to get him to the reading of the names, a page-long list of the dignitaries present that includes virtually the entire black leadership of the United States since the '60s—John Lewis; Jesse Jackson; old standbys such as Benjamin Hooks; the children of Martin Luther King, Jr.; Ralph Abernathy's wife, Juanita; NAACP chairman Julian Bond. Also in attendance are Nancy Pelosi and every liberal Democrat who could get through the gates, along with a few stalwart Republicans, such as obstreperous Judiciary chairman Arlen Specter.

Bush, seeing the end in sight, seems to ease up, winging it, thanking the Washington mayor for coming, and quipping, "Everything is fine in the neighborhood. I appreciate it."

He gets a laugh, a victory, and four minutes later he signs the reauthorization. Bush said he'd do anything possible for his Republican brethren. What's more, facing a Democratic Congress, at this point, would all but end his presidency and, he believes, make America less safe. He hopes this signing helps matters, but he's skeptical. He knows this crowd will never forgive him for the way black voters were disenfranchised in Florida in 2000, or for the way he's snubbed them since, or for Katrina. But what did it really matter? The civil rights community is old—just look at them—from a time when it was all about winning rights. Now it's about equal opportunity, a slippery standard. And since 9/11, race—like so many other issues—has been eclipsed by more urgent matters.

But the really urgent matter is how fast he can get away from all these doubters—angry black doubters, at that—and back into his safe house. Bush often tells his senior staff that his schedule matters, that he's prompt and decides, in advance, how much of his time a particular task or meeting should command. "My schedule," Bush once said to Colin Powell, "is my way of sending a message about what I think is important or not." Powell, of course, met with the president privately only a handful of times during his five-year tenure. In this case, Bush walks briskly from the ceremony at 9:52. His speech, the signing of the reauthorization, and brief cordialities take exactly sixteen minutes.

———————

A parallel conversation about race is occurring, meanwhile, near the Alexander Hamilton statue.

"Reggie? It's me, Usman."

"Yo, Khosa. Where are you, man? The staff meeting is starting, like, right now."

"Listen, Reggie. I've been detained down on the street, right in front of the White House, I mean in front of Treasury. They think . . . you know . . . well, me being from Pakistan and all."

The receptionist at Barnes Richardson, Reggie Mc-Fadgen is a large jocular African American, thirty-three, who grew up in Washington, went to the tough D.C. schools, and once sang with the hip-hop group Salt 'n' Pepa. He and Usman have become close friends over the past two years; they go out often. Reggie, tough and streetwise, is protective of "Khosa." They call each other "my brother from another mother."

Like many educated foreigners, Usman happens to know quite a bit about American history; after all, the history of the United States, the world's most powerful country, is taught in most accredited schools overseas. Special attention is generally accorded to slavery and to the current state of the black American. The long saga is viewed as America's original hearts-and-minds struggle, and, from afar, the nation's character is often assessed, surprisingly, through this lens. What's clear, and oddly moving, is that Usman is crazy proud that he and his black friend call each other "brother." It makes him feel like an American-in-the-making, free, by virtue of his Pakistani heritage, to improvise some solutions to the country's long-standing dilemmas of race.

At this moment, though, Reggie is feeling that he possesses the pertinent expertise about what Usman is facing. His friend is being profiled.

"Khosa!" he shouts into the phone. "Don't you move or say a word to any officer of any fucking kind. I AM COMING DOWN TO KICK SOME ASS!"

Oh God. "No, Reggie, please. Trust me. That'll just make things worse. I'm in enough trouble. Just go to my top desk drawer and get my passport and put it somewhere safe. I may need you to bring it somewhere. I'll try to call you as things progress. Okay?"

A second later Reggie is racing through Barnes Richardson like a town crier. Usman has been detained! Right in front of Treasury! Work stops. People pour from their offices toward the suite of the boss, Matt McGrath. They crowd around the wide window in his office, trying to catch a glimpse of what's happening on the street below.

Ten stories down, things are progressing quickly. A black SUV screeches up to Alexander Hamilton. Two men in dark suits get out.

"Usman Khosa?" one of them says, a tall, neat man, midthirties, with dirty-blond hair. "Get in the car."

"No way. I'm not getting in that car," Usman says, surprising himself. He feels like he's going to vomit. These are the bosses of the uniformed guys. He's read about Guantánamo. If he gets in that car, he may never be seen again.

"Mr. Khosa, it's not a question," Dirty Blond says, emphatically. "Get in the car."

Usman backs away. Assesses his options. "Okay, okay, I'll get in . . . just as long as I can make a few quick phone calls first. Then, I promise, I'll go with you."

Dirty Blond nods, and then huddles with his partner, a short, wide, ethnic Italian-looking guy.

Usman calls the Pakistani embassy, tells them he's being taken into custody. They take down his information and his present location. There's nothing they can do. He calls his friend Zarar, a guy who runs a large network of Pakistani young professionals and who knows people. "If I don't call you in two hours, Zarar, call someone. I don't know who. Anyone!" Then, he pauses for a moment, a moment to wince, before he calls Pakistan. Tariq Khosa is currently the second-ranking law enforcement official in Punjab Province, home to nearly half of the country's population, along with Lahore, the nation's cultural hub, and Pakistan's capital, Islamabad. Usman is sure his father currently knows people in law enforcement in America. He must. All he gets is voice mail. He tries to compose himself: "Dad, I'm in trouble. I'm being arrested by the Secret Service. I have to get into a car, a black SUV, in front of the White House. I don't want you to worry, but if you don't hear from me in a few hours, um, see, maybe, if you could call someone." He pauses. "And please, don't tell Mom."

While the old guard from the glory days of Montgomery bus boycotts and King's speech about American dreams slowly disperses—the ink drying on a renewal of their greatest legal victory—an unscripted ceremony of fear and hope and race is unfolding on a nearby sidewalk.

In it, a young Pakistani man, whom terrorists have succeeded in turning into a racial suspect, steps into a black SUV. Ten stories up, a consulting firm that looks and feels like the great promise of America's future—every major race represented, competing fiercely, hoping for bonuses and playing softball on Saturdays—watches, in horror, faces of every color pressed against a wide plate-glass window. One of their own is down there. But is he one of them? Of course he is. About the best analyst they've got, a gentle kid, smart as a whip, works like a stevedore and hopes for things worth hoping for. But then again he's being taken away by grim authorities of the U.S. government and maybe there's a chance—a slim chance—that there's some good reason for it; that, maybe, they don't *really* know him.

The black SUV's doors slam shut. "This is so wrong," Reggie says, almost to himself. Matt McGrath, next to him, snaps to attention: "We should call anyone that anyone knows in the government." A dozen people run to the phones.

The SUV, meanwhile, makes a U-turn and drives back toward 1600 Pennsylvania Avenue, stopping at the far entrance. Usman is trundled from the SUV, escorted through the West Gate, and onto the manicured grounds. No one speaks as the agents walk him behind the gate's security station, down a stairwell, along an underground passage, and into a room—a cement-walled box with a table, two chairs, a hanging light with a bare bulb, and a mounted video camera.

Even after all the astonishing turns of the past hour, Usman can't quite believe there's actually an interrogation room beneath the White House, dark and dank and horrific. He mops the perspiration from his brow.

Real sweat. It's no dream.

Aboveground, the bounce has returned to George Bush's step.

Romanians. He likes the Romanians, and he likes their president, Traian Basescu. And what's not to like? His aim is to please the United States, to be a member of the club. That means both the European Union—which Romania is hoping to become a full member of soon—and the coalition of countries aligned, without a doubt, behind the United States.

They sit in the wing chairs near the fireplace and chat, the two of them. Basescu has a visa issue he wants resolved. Bush says he'll get on it, open things up for sure. An early member of the coalition of the willing, the Romanians have stuck it out. They're the only country in their region with soldiers in both Afghanistan and Iraq.

After a half hour of chat, Bush rises. "Let's bring in the jackals."

Reporters file into the Oval Office.

Bush reads a brief statement, written before the meeting, of course, about what they've just discussed, and then Basecsu reads his version of "we are good friends as are our nations."

Bush first goes to Jennifer Loven of the Associated Press, who dives right into the morning's fare, that Israel's Justice minister has said that the lack of a cease-fire call from the international community at the Rome conference—a result of U.S. opposition—gives Israel the "green light" to push further, and that a top Israeli general is saying that fighting will continue for a few weeks. "Is your administration okay with these things?"

Bush sighs. Here we go. "I believe that, as Condi said yesterday, the Middle East is littered with agreements that just didn't work. And now is the time to address the root cause of the problem. And the root cause of the problem is terrorist groups trying to stop the advance of democracies." It's a stretch, and he knows it. Hezbollah provides social services and protection to a significant segment of Lebanon. They're settled. They have offices, weapons depots, bank accounts. They're more like an unauthorized government. And he goes with that, moving from terrorism to issues of electoral legitimacy.

"I view this as a clash of forms of government," he says to the reporters. "I see people who can't stand the thought of democracy taking hold in parts of—in the Middle East. And as democracy begins to advance, they use terrorist tactics to stop it. . . . But our objective is to make sure those who use terrorist tactics are not rewarded." Don't reward bad behavior—that's what Cheney often says—but Bush, at this point, knows this kind of global experiment in behavior modification is like running a calculus equation with fast-changing variables. Surprises every minute.

He pauses, exasperated. "Want to ask somebody from the Romanian press." He gets a Romanian question and talks about visas and the Black Sea, all the stuff in his briefing book. Then the next question—from Steve Holland of Reuters—mentions a tape released yesterday by Zawahiri, urging Muslims to fight and become martyrs.

Bush seems to lift. He wets his lips, and his head does that funny forward cock. He can connect it all—the political, the theoretical—through this one. He can make it personal. Him versus Zawahiri . . .

"My answer is, I'm not surprised people who use terrorist tactics would start speaking out. It doesn't surprise me. I am—Zawahiri's attitude about life is that there shouldn't be free societies. And he believes that people ought to use terrorist tactics, the killing of innocent people to achieve his objective. And so I'm not surprised he feels like he needs to lend his voice to terrorist activities that are trying to prevent democracies from moving forward. . . . You know, here's a fellow who is in a remote region of the world putting out statements basically encouraging people to use terrorist tactics to kill innocent people to achieve political objectives. And the United States of America stands strong against Mr. Zawahiri and his types."

A floor below, Usman is being grilled about whether he's in league with "Mr. Zawahiri and his types." What that means, as one hour moves into the next, is he's forced to walk along the knife's edge of determining whether anything about his life—anything—is suspicious.

The question is if anything doesn't "quite add up," as Dirty Blond tells him. He's the bad cop, taking notes, one legal pad after another.

"This might sound crazy," says the Italian, the good cop, "but we need to know the name of everyone you know in America pretty much. Every friend from college, girlfriend, professor you hung out with. People at work. Everyone."

Usman, all but suffocating under his status as terror suspect, now must spread the virus to everyone he knows.

He freezes. Says nothing for a moment. Italian smiles and leans in. "You live with anyone here in Washington. Roommates or something? Start with them, and phone numbers."

So Usman gives up his ecumenical cell, Linas and

David. And their numbers. Dirty Blond silently marks it all down.

The names build, one after another. Usman pauses, name by name. His whole life in America, *everyone he's met.* "Is it really that important that everyone is included in all this?"

"Say you leave someone out," Dirty Blond interjects, "and they turn out to be someone who is of interest to us. Well, that's a bad thing—bad for you."

On he goes, an hour passes. He thinks about his father, who has interrogated al Qaeda suspects in Pakistan. What would Dad tell him to do? Just be completely honest. But the truth can be complicated, and an area of confusion can so easily create suspicion. Then, God forbid.

Another agent arrives with an atlas.

"I don't know much about Pakistan," Italian says. "Show me where you're from in Pakistan."

Usman opens the atlas to Pakistan. He points to Lahore. Dirty Blond's up, looking over his shoulder. "That's close to the areas where there've been some problems, isn't it?"

Usman looks up, then down. Yes, on the map the wide span of Pakistan seems to shrink. Lahore looks as if it's right beside Peshawar and the edge of the tribal areas.

"It's farther than it looks—on the map, I mean. Like five hours' drive."

Then the questions come in waves.

"What do you think of Islamic fundamentalism?"

"What's your opinion about Osama bin Laden?"

"Define for me the word *jihad*."

"You seem to know a lot about it."

It's like there's a trap in every question, and each one of his responses seems to ensnare him. How much knowledge of these things is good to have? How much adds to suspicion?

"I know some things, I guess, because my father is, um, a police chief in Pakistan."

That should help. But then Dirty Blond seems to infer that Pakistani authorities have been infiltrated by radicals. "Is he a radical?"

"No . . ." Usman sputters. "He loves America. We even came here for a year when I was a kid. We lived in Seattle. He was funded for a year in America as a Hubert Humphrey Fellow."

Dirty Blond looks at him, affectless. "And then you went back. So you learned a lot about America, how things work here, and then went back to Pakistan."

"Yes, I guess so. I mean, I was in first grade."

More notes are jotted on the pad.

Usman looks, pleadingly, at Italian. "Can you at

least tell me why I was picked up? Was it just the way I look?"

"It was also what you were doing, fooling with your iPod as the motorcade passed. That didn't help. You understand how bombs are detonated by remote devices?"

No, no, I'm not going there.

"Um, no, not really."

Dirty Blond: "You're from Pakistan and you don't know how things are detonated with radios or cell phones or PDAs?" He seems angry.

"Well, I guess—yes, I do." A lie.

More notes on Dirty Blond's pad. *I'll never see my mother again.*

Italian, smiling, asks, "Why did you think you were stopped?"

Usman pauses. He's drenched in sweat. Hasn't eaten yet today. He was going to get a muffin this morning near the office. He's been underground for hours. Has to be at least midafternoon. He tries to answer with only the facts he's sure of.

"Well, a guy in a uniform stopped me."

"What kind of uniform?"

"Secret Service, or something else." He can't pull it up.

"You don't remember? It just happened."

"No, no I can't." This draws arched brows from both agents.

They ask him a long list of medical questions. Has he ever been treated for depression? Paranoia? Bipolar disorder? The list goes on.

Dirty Blond produces a form, a medical release form.

He pushes it across the table. "This is so we can access all your medical records, to check everything you've just said."

Usman is thoroughly beaten. He just nods, silently, and signs.

Both agents rise, Dirty Blond in the lead. "If anything you told us today doesn't check out, we know where to find you. And we will find you."

Usman nods, looking down. "Let's go," Dirty Blond orders.

As George W. Bush is returning to the White House—having just finished a rousing speech about the "miracle of the U.S. economy" to manufacturers gathered at the Grand Hyatt Washington—Usman Khosa is leaving it.

Dirty Blond leads him through the security gate and onto the sidewalk. He seems oddly jocular, assuming Italian's friendly posture. They walk along, as though they'll be staying together.

"So I think I've got you," he says, like it's a joke. "Why are you wearing shorts and a T-shirt if you're really on your way to an office, right over there?"

He stops. Usman stops. They're face-to-face, or rather forehead-to-chin. Usman measures every word. "Because my suit for today is at the cleaners near the office. I was going to pick it up."

Dirty Blond stops, calculating the response. It's clear he was saving this one—saving it until Usman had let down his guard. "So what's the name of the cleaners?"

"Oh God, I really just don't know."

"Okay, you're free to go. But I'll be checking out everything you've told us. And, at this point, we know everything about you and where to find you." He turns and walks briskly back toward the White House. Usman is left standing on the sidewalk.

And so two characters—a president, chosen to represent America to the world, and a Muslim striver, who's acutely representative of a struggle now turning history's wheel—watch the day's shadows lengthen.

One reads through security briefings, preparing for his meeting the next morning with Tony Blair when he plans to talk frankly about this airline plot.

The other walks across the street to Lafayette Park, sits in the grass, and weeps.

After a time—could be five minutes, or twenty-five—he calls his friend. "Thank God, Usman, thank merciful God," Reggie says. "Look, come over to the office. Everyone here's been worried sick." Usman says he will, but first he calls his other friend, Zarar. He calls his father, leaves another message. He calls Linas at his office, and David at his. He tells them they may be getting calls from the Secret Service. He sits some more, wondering how many people of all those he mentioned will now be called by the FBI, or the Secret Service, or Homeland Security, or CIA, or whoever the hell follows up on this sort of thing. Will some of them get tripped up in the questioning? Should he call them all and alert them, or would that constitute "suspicious behavior"?

These questions distress him for a half hour. He begins to wander over toward Alexander Hamilton. Looks up at the pointed, patrician chin, the aquiline nose, and then crosses Fifteenth Street. When he gets off the elevator on ten, Reggie leaps from the reception desk and throws his arms around him. He's talking excitedly as others start to gather in the carpeted reception area. The whole gang seems to be moving, carrying Usman along, toward the conference room.

Usman sees something on the long mahogany table. The crowd parts. It's a cake, a sheet cake the size of a car door. Linda the office manager ran to a bakery across the street and got it once Reggie trumpeted the good news. Written in blue on the white icing are the words welcome home usman.

What a strange and surprising country this is, Usman Khosa thinks, lifting the first slice to his lips. Heaven and hell, and then we eat cake. His mind reels, as his heart searches the faces—his friends, his colleagues, this makeshift American family—aching to know if he's still one of them.

Chapter 2
Takeoffs and Landings

Events move forward swiftly and in secret. These are the operational trademarks of the era of Bush and Cheney. Keep the information loop small, a tight circle of trusted people. The actual decision is often made by just the two men, alone in a room—the Oval Office, the video room at Camp David, the small dining room on the second floor of the White House, where they lunch.

Once the *why* is decided, keep discussion of the *how*—how it will be done—as tightly controlled as possible. Just like running an intelligence operation, people will sometimes see part of the equation—this was ordered, that was done—but not the whole.

For complete context you need to pull back a distance and gauge, first, the traditional dilemmas of the so-called political mandate. Politicians have always felt its pull—survival's tug—and have had to decide, day to day, where or whether their political self-interest coincides with the broader interests of the nation. Their reelection, or place in history, often depends on their choices along this axis.

The oddity, however, of the modern conflict between governments and terrorists, conducted largely in secret, is that this tendentious line can be drawn in the privacy of the White House, invisible to anyone except those who need to know.

Along with its penchant for secrecy, a feature of the Bush presidency, clear at this point, six years along, has been a sense of messianic purpose that makes the national interest almost indistinguishable from the political interests of the president. What triggers action, thereby, is often Bush's simple dissatisfaction, the garden-variety frustration of not getting one's way.

And the meeting with Blair on the morning of July 28 happens not to go well. When Bush brings up the airliner plot and expresses his desire to snap the trap shut, Blair is unmoved. He says he's quite clear about the position of his people—Eliza Manningham-Buller, the imperious head of MI5, domestic intelligence; and John Scarlett, the chief of MI6, the foreign service. They seem to have prepared Blair for just this kind of push from the Americans. It's not just that this is a UK operation, Blair says, and that nearly two thousand British operatives have been working it for nearly a year. It's also that if they're patient, at some point they'll be "at the ready" when the plotters seek "green light" approval from al Qaeda's chiefs. It's too large a plot for them not to. After all, it was U.S. intelligence, Blair points out, that discovered how Zawahiri called off the New York City cyanide-in-the-subway plot in 2003 for "something better." With a plot as big as what is slowly developing in London, the terrorists will surely seek permission to move forward, and when they do, Blair asserts, we can run the thread right into Zawahiri's beard.

By then, it's time for their joint press conference. Bush is not going to get anything. After Blair leaves, Bush tells Cheney about his dissatisfaction. Cheney receives the message clearly, as he's received many others over the years. He knows how concerned Bush is about the coming midterm elections, that Congress will go to the Democrats and the rest of his administration will be a wash of gridlock and recrimination. Cheney, of course, is similarly concerned. And now Bush wants something done. No one needs to create any memos, which some historian might dig up. In a private, two-man exchange there is an understanding. This is the way their relationship works, especially on the most sensitive matters that Bush will want to deny if he's ever confronted.

This sort of deniability is the product of the arrangement Bush and Cheney slipped into in the first few months of the administration. That's when Bush told Cheney that he had to "step back" in large meetings when they were together, like those at the NSC, because people were addressing and deferring to Cheney. Cheney said he understood, that he'd mostly just take notes at the big tables and then he and Bush would meet privately, frequently, to discuss options and action. This gave Cheney the structural latitude to carry forward his complex strategies, developed over decades, for how to protect a president. After the searing experience of being in the Nixon White House, Cheney developed a view that the failure of Watergate was not the break-in, or even the cover-up, but the way the president had, in essence, been over-briefed. There were certain things a president shouldn't know—things that could be illegal, disruptive to key foreign relationships, or humiliating to the executive. The key was a signaling system, where the president made his wishes broadly known to a sufficiently powerful deputy who could take it from there. If an investigation ensued, or a foreign leader cried foul, the president could shrug. This was never anything *he'd* authorized. The whole point of Cheney's model is to make a president *less accountable* for his actions. Cheney's view is that accountability—a bedrock feature of representative democracy—is not, in every case, a virtue.

So over the next week, as Bush packs up and arrives in Crawford, Cheney makes provisions. It's all very tightly held. No one can know what's under way—no one, certainly, in the foreign policy establishment and not even most top officials in the intelligence community.

If this were run on a split screen, one image would be a man slipping into Islamabad in darkness in early August. The man is Jose Rodriquez, the director of operations at CIA, the agency's number four official, and the head of all clandestine operations and CIA stations around the world. He was moved into that position in 2005 by Porter Goss, as part of Goss's mission to rein in the renegade agency and make it more attentive to filling the needs of its "customer," the White House. Outside of Cheney's office, virtually no one in America or abroad knows that Rodriquez has been dispatched to Islamabad. It is not, however, Pakistan's powerful Directorate of Inter-Services Intelligence, or ISI, that Rodriguez and Cheney are concerned about. It's that British intelligence will discover Rodriguez's presence through their sources in Pakistan or those in the U.S. intelligence community.

His mission is to secretly pass information to a selected Pakistani intelligence official, who will summarily arrest one Rashid Rauf, the Pakistani contact for the British airline plotters. Rodriguez then has to get out of the country undetected.

He manages this, and Rauf is quietly apprehended. As a whisper of the arrest spreads to a few top officials in British intelligence—first in Islamabad and then in London—they curse, throw ashtrays, and scream bloody murder. They know the Pakistanis would never have moved on Rauf without first checking with them. And God knows the Brits didn't the give the order.

This was their investigation. They might have made arrests in a week, or in a month. Clearly the hijackers were moving into some next stage of planning, but many inside of the British intelligence and law enforcement communities were beginning to suspect it was more experimentation than anything resembling a "dry run." If so, taking them down too early would leave investigators with insufficient evidence to effectively prosecute. The British model is, after all, to be patient, gather sufficient evidence to try terror suspects in open court, and get long prison terms, treating it all as a criminal matter rather than a historic—and terrorist-glamorizing—clash of power and ideology. As for Rashid Rauf, the British had even more specific plans. He was wanted for murder in the UK. The Brits were preparing a case, with plenty of evidence, for the Pakistani police to arrest him and have him extradited to England for trial, just like any murderer on the lam. Instead, he gets picked up by the notorious ISI, where he'll be either tortured or feted—depending on ISI's complicated views of the matter—and rendered unsuitable for public trial in the UK or anywhere else.

His arrest lights a fuse that will swiftly implode their entire investigation. Top U.S. officials are perplexed. DHS intelligence chief Charlie Allen, a CIA legend who helped uncover the Iran-Contra affair, predicted the Iraqi invasion of Kuwait in 1991, and was ahead of the curve on al Qaeda in 1998, tells the Brits that he and DHS are surprised as well. "Jose had to get the green light from upstairs, but no one here was briefed," a senior intelligence official recounted later. "We were all shocked." No one, beyond a half-dozen people in the entire U.S. government, several of them among Cheney's national security team, knew who was responsible.

But not everyone is upset. Bush suddenly is in a very good mood. He'd seemed a bit glum when he first arrived in Crawford, aides said. But by Tuesday, August 8, he is notably ebullient, spirits high, watching the thermometer with glee. It's turning into a Central Texas scorcher. Ninety-seven degrees by late morning. Now ninety-nine! Just after 1:00 p.m., as the temperature passes one hundred, he starts rousting up staffers: tryouts, immediately, for the president's 100-Degree Club. The rules? Run three miles when the thermometer hits triple digits. The coveted prize: a gray Under Armour T-shirt with the insignia "The President's 100-Degree Club" framed by a Texas Star. You also get a photo with Bush, commemorating membership in the club. Eighteen White House aides, among them the press secretary, Tony Snow, line up and set off. Bush, who blew out his knee jogging in 2003, pedals around them on his bike, mocking the leaders, taunting the stragglers. He's giddy. The ones who make it the three miles, including Snow, gasp toward the finish line, purple and breathless. The rest, also called losers, get no shirts. Then, pumped up, Bush hits the phones, making preparations for some political jaunts over the coming days to assist beleaguered Republican candidates. He knows he'll be able to deliver something for them, an August surprise, something very good for the Protectors of America.

On Wednesday, August 9, Cheney decides he wants to do a surprise press briefing from his vacation home in the Wyoming hills. This is unusual. Reporters sign on. The previous evening, antiwar candidate Ned Lamont defeated Joe Lieberman in Connecticut's Democratic primary. Cheney is notably strident, saying that Lamont's victory will encourage "al Qaeda types."

"And when we see the Democratic Party reject one of its own," Cheney continues, referring to Lieberman, "a man they selected to be their vice presidential nominee just a few short years ago, it would seem to say a lot about the state the party is in today if that's becoming the dominant view of the Democratic Party, the basic, fundamental notion that somehow we can retreat behind our oceans and not be actively engaged in this conflict and be safe here at home, which clearly we know we won't—we can't be." He goes on to describe this as a sort of "pre-9/11 mind set, in terms of how we deal with the world we live in." Now he just has to sit in his vacation home and wait for events to catch up with his words.

Of course, everything becomes clear in the early morning hours of August 10. An associate of Rauf's in Pakistan alerted one of the British plotters that their man in Pakistan had been arrested—a conversation that is picked up by electronic surveillance shared by the UK and the United States. As the plotters panic, British police slip into high gear. They race across metropolitan London, rounding up more than twenty suspects in a few hours, shutting down a yearlong operation in what can only be called a frenzy. The most knowledgeable

British anti-terrorism officials are the most outraged. Before dawn breaks in the UK, they're already assessing the damage from what one calls a "forced, foolish hastiness." The haul from a suburban "safe house" shows that the plotters had begun to experiment with ways to get the explosives aboard the airliners. In a general way, they'd started to examine the pattern of flight schedules to the United States. In other words, early planning stages—the most valuable time for British agents to sit and wait, tug the net gently and watch for the arrival of "operational readiness," the moment the plotters would seek permission from on high.

But by the time panicked British police are kicking down doors, the White House already has a media strategy set to leverage news of the thwarted attack, "the worst since 9/11." All that's left to do is wait a few hours until sunrise, when the arrests will hit the U.S. news cycles and the president and vice president can register surprise about how right they've been all along, *about everything.*

While America sleeps, Naeem Muhsiny is walking through the Frankfurt Airport, thinking of Moses.

And he hates that. Religious allegory, *there's no getting away from it!* He views religion, all religions, as "an illusion, an opiate," though he won't tell that to just anyone or how that conclusion grew during his youth in an Afghan madrassa and deepened across the years he endured the sectarian strife of his country and neighboring Pakistan.

No, such candor doesn't fit with Naeem's current obligations, shepherding Afghanistan's brightest teenagers to America for an extraordinary experiment in cultural connection, in the possible. The mission: pluck forty kids from across the war-torn Islamic nation, teach them everything that can be learned about America in a month of orientation, find families in the United States to host them, and then manage a journey across a dozen time zones and several centuries to deposit each unwitting youngster into the living room of some volunteer family and the bustling halls of an American high school. In a year they'll return home— if they can survive that long.

"Pair off in twos!" he shouts above the airport din. "This is a buddy system. STAY WITH YOUR BUDDY!"

He glares as they hustle into a double helix, then two neat columns.

"Fine," he grumbles. "Now follow Mary."

Mary is a tiny gray-haired Indian lady, Bombay-born but now a German citizen, whose main job is to guide confused exchange students across the airport on behalf of several German travel agencies. She raises a minuscule fist and begins marching as, two by two, the train of dizzy, half-smiling teens lurches forward.

They are the cream of the select. Twenty-five hundred Afghan youngsters between ages fourteen and seventeen applied eight months ago to something called the American Councils for International Education and took a test. Seven hundred of the top scorers then took another test and wrote an essay. Three hundred of those were asked to complete a long array of essays and sit for an interview. Forty made it, winning a trip to the United States care of American Councils, a three-decades-old organization, funded by everything from the State Department to the World Bank to George Soros.

Naeem, hardened and pragmatic at thirty-two, winnowed through all the applications with his boss in America. That's where Naeem lives with his wife, deep in the Virginia hills, about two hours from D.C. in a barely affordable house that expands "exurb" to its theoretical limit. He knew these kids first through their photos and the reports from colleagues who interviewed them in Afghanistan, and then got to know them better during a month at an abandoned ski resort in Kyrgyzstan, where he assisted with their orientation and taught them the strange customs of America.

Today's group of twenty is the first shift—kids whose

first days at their American high schools come later this month: August 2006. In a few weeks, when the September group will cross over, the twenty kids now snaking through Frankfurt will already be wandering across a landscape they have dreamed mightily about—dreams jammed with harebrained assumptions, ferocious yearning, and plenty of global economy product placement.

Mary halts the train in an airport lounge as Naeem, bringing up the rear, circles around to huddle with her. "So how've you been . . . how's your wife . . . where'd you get that lovely leather jacket?" Mary is chatty, a busybody. This is a lark for her, a part-time job. Naeem can't believe she's making small talk at a time like this—a moment he's worked nearly a year to create. He reminds himself to breathe. "Yes, great, Mary. Wonderful. How about food?" The travel agency that three months ago booked the twenty-one low-cost tickets—twenty students plus Naeem—in a block purchase includes one lunch per student in its package.

"Everyone gets *Filet-O-Fish* sandwiches!" Naeem barks, unable to suppress a smile, as titters run through the group. "Any problem with that!" Silence. The fish sandwiches, launched by McDonald's in 1962 to meet the Catholic prohibition against eating meat on Fridays, are the only fast-food choice.

The lunch order, like so much that transpires, is the result of fastidious cross-cultural planning, mostly by Naeem. The students, who must have a workable grasp of English, are instructed to speak neither Farsi nor its Afghan cousin, Dari, which might attract unwanted attention in the post-9/11 era. All commands from Naeem are in the language they'll be using in America. As for comestibles, only the meat of animals slaughtered in accordance with Koranic principles is halal, a designation almost identical to that of kosher. So the holy twenty—future leaders of global harmony—eat fish, and French fries, and suck down Cokes, ending this small s sacrament with the gentle folding of licked-clean McDonald's wrappers for storage in their travel bags. One dream met, and conquered.

Greasy and full-bellied, they file wide-eyed past airport shops and across concourses that gleam with shaped metals and glossy polymers that are as foreign to their eyes as the Mars surface would be to a Chicago bus driver.

Naeem checks his watch. An hour until the departure of the Lufthansa-United flight to America.

"Boarding passes and visas ready. Hold them high!" he shouts, "Now, single file." Naeem counts about thirty passengers ahead of his group in the queue to pass through the security checkpoint. He does a head count as his kids, all in sky blue T-shirts emblazoned with yes, for the program's official name, Youth Exchange and Study, line up straight and solemn—twenty strong. He checks his watch again. Almost there.

He hears something to his left. The line stretches alongside a bar, a casino bar, with slot machines and a sixty-inch flat-screen TV.

Naeem motions to Mary—he points to his left—and begins to drift over.

The bar is crowded at 10:03 a.m., with people milling about. Naeem looks up at the big screen, at the blue studio backdrops of CNN International.

BREAKING NEWS . . . "The most significant plot since the 9/11 attacks was foiled this morning by British police . . . bombs on multiple aircraft due to explode over the continental United States . . . arrests early this morning in and around London . . . all liquids and gels are banned on flights in Europe and America . . ."

A tiny hand grabs his sleeve. It's Mary, and with her is a tall, sandy-haired man north of six feet, in a snappy United Airlines uniform.

"I think we're going to have to pull your group from the line for a special security screening," he says in German-accented English. Naeem, a wide-shouldered five-foot-six, looks up at the man's gleaming teeth.

He can't feel his feet, but they're moving. "Okay,

everyone out of line, follow me," he says to his twenty charges, trying to make it seem like no big deal. And then they're walking down a corridor toward descending steps, led, now, by the man in uniform.

Mary's panicking. "Oh God, there's no chance we're making this flight," she says in a stage whisper, skipping to match Naeem's stride. "I mean, we'll have to get a hotel. Okay. Or send them home. How will we buy twenty tickets? *They can't go on three different flights!*"

"Just keep cool, Mary. You know what the word *cool* means?"

Of course not. "No one's going to America today, or tomorrow," she says to herself. "No, no, no, no."

Ten minutes later, they're in a netherworld of small gates for small flights going to nameless European cities, and then they're herded into an empty gate area with a private screening room for bag searches and pat-downs.

Two large uniformed men arrive to start the checks; Naeem's bearings return. "You can't have men pat down the girls," he says, putting on his angry mullah look. "You'll need a woman."

The toothy, sandy-haired official seems to sigh, *Muslims*. "Fine. I'll return."

He doesn't. Five minutes later he's replaced by a petite blond woman, late twenties, upturned nose, blue eyes, hair pulled back, a United uniform fitting snugly. Naeem finds himself resenting Western ideals of beauty. Maybe he'll never see another Caucasian. That could work.

"Listen to me," he bears down on the woman, grateful that she's short. "Our plane leaves in forty-six minutes. There are twenty of them. This has to happen very quickly."

Blue eyes meet brown. "We'll do what we can, sir."

After twenty minutes, four kids are through. They have to answer twenty questions, unpack everything, and dump precious pastes and fluids, colognes and toothpastes and shampoos they've bought with scarce "travel money" during orientation—a month when they were told repeatedly how crazy clean Americans are.

Naeem, pacing outside the private screening room, looks over at Mary, who's talking to herself. The cleared students sit quietly, afraid to speak, holding their bags with blue Lufthansa-United stickers, denoting a properly checked carry-on. Naeem ducks his head into the screening room.

"We're going as fast as we can," the blonde says, all business. "I have no control over procedures."

Twelve minutes pass. The plane leaves in fourteen

minutes. Ten kids have made it through. A United agent bounds down the stairwell and is running toward them. "Look," she stops, panting. "The children who have passed the security check *have* to get on the plane. The plane is *leaving*."

She meets Moses. "Don't moooove," Naeem rumbles. He marches to the screening room, throws open the door.

The blonde snaps around. "Your friend out here says that the plane is leaving only for my kids who've been cleared. Let me explain, THAT IS NOT GOING TO HAPPEN. WE GO TOGETHER OR NOT AT ALL."

This man from Afghanistan and this woman from Germany are now standing, alone, in a place beyond procedures and supposed-tos, beyond race and suspicion, and there are children involved.

She looks around his shoulder, out at the kids still lined up for pat-downs, clutching their bags. Blue eyes meet brown.

"Okay, let's get them all through, quickly." She motions to her assistant. The kids crowd in, as the magic blue "United" stickers are slapped on their bags.

"Thank you," Naeem whispers.

"Run," she says.

And on a day the world awakens to news of narrowly averted disaster, twenty Muslim teenagers in their blue shirts—short sleeves for the boys, long for the girls, whose arms and heads are covered—run for their lives.

Naeem awakes to an announcement—"Prepare for landing in Washington"—and leans forward to hear two students, girls, talking in the seat in front of him.

"I don't see it," one says to the other, as she looks out at the gentle Virginia hills surrounding Dulles Airport.

"What?"

"You know, the building. The place where Bush lives."

She searches for the words in English, then slips in forbidden Farsi.

"Qasre Safid." *The White Palace.*

The occupant of Qasre Safid is, at this moment, preparing to land halfway across the continent, feeling a satisfying swell of anticipation. Air Force One touches down at Austin Straubel International Airport in Green Bay, Wisconsin, and a wartime leader, Bush's favorite self-designation, purposefully disembarks to a tarmac podium with remarks he's been wanting to make since late July.

"The recent arrests that our fellow citizens are now

learning about are a stark reminder that this nation is at war with Islamic fascists who will use any means to destroy those of us who love freedom, to hurt our nation," he says. He is alone at the podium—the Republican candidates he's here to visit are invisible, pushed away to his far left. The line of sight from the press riser to the podium places the word *UNITED*, painted on the side of Air Force One, directly above Bush's head. Nearby, a satellite truck hums. "I want to thank the government of Tony Blair and officials in the United Kingdom for their good work in busting this plot. . . . Cooperation between U.K. and U.S. authorities and officials was solid. . . . The American people need to know we live in a dangerous world, but our government will do everything we can to protect our people from those dangers. Thank you."

It is his first public comment since the news of the foiled attacks broke. As those words circle the globe, leading every news report on the planet, Bush stops by a sheet metal plant to mug with the workers, put on the hardhat, and offer a short factory-floor economic speech, "to remind the American people that by cutting taxes on small businesses, it encourages small businesses to grow." Then he and five hundred supporters lunch at a $1,000-a-plate fund-raiser for Wisconsin State Assembly speaker John Gard, a Republican running for an open congressional seat.

In all, a perfect day—a day to amplify, as part of his official duties, that Americans should be very afraid but also very grateful that his government is doing everything possible to avert catastrophe, and, in passing, to mention what he, and tax cuts, have done for the economy.

The kind of day to lift a president's spirits—a day when he can briefly own both fear and hope. Mission accomplished.

Ann Petrila watches Bush on television in her Denver home. Hates the man with a profound, volcanic, and deeply satisfying rage. The tarmac statement from Green Bay is replayed endlessly among bits of evolving news about the thwarted attacks. Saturation coverage, round-the-clock. The phone rings in the kitchen.

"The kids have been held up at security in Washington and missed their connecting flights," the caller—someone from the American Councils—tells her. "At least, that's the report we've received. We'll call as soon as we get any more information."

She hangs up and heads for the basement, a wood-paneled cave that her son, Ben, will be sharing with their Afghani exchange student.

He's pushing thumbtacks through a small poster with red, white, and blue stripes drawn around the words welcome to america, mohammad.

"I can't believe I'm putting a sign like this up in my room," he tells her. "I think I've lost my mind."

She laughs. This three-bedroom house—a classic, lovingly preserved Sears Craftsman, circa 1911, with a sun porch on the front and oak beams—is the progressive home of Ann; her husband, Michael; and their son, Ben—a fallen Catholic, a Buddhist (by choice), and an atheist, respectively.

Ann goes out the screen door to the garden, plops down alongside a row of tomatoes, and starts weeding. She is a professor of social work at the University of Denver's Graduate School. Michael is a psychologist at an academy for troubled youth. By training and inclination, Ann is sensitive and, when need be, quite candid. Like many Americans, she believes in getting everything on the table and talking it out. As she tears at the dandelions, she thinks of this poor child, a seventeen-year-old boy named Mohammad, and what it must be like to come from Afghanistan to America and end up in some interrogation room in an airport.

This was a lark, this idea of having an exchange student, which took on a life of its own and, as she digs deep for the dandelion roots, she thinks of sitting with Ben and Michael at the dining room table back in May with folders of possible students. She remembers there was a boy from Serbia, a religious Christian.

"Yuck," Ben said. "A fundamentalist Christian, no way."

Ann laughed. "Well, a Muslim from the bowels of Afghanistan. He's going to be religious, too, Ben."

"I know—but at least he's not some Holy Roller."

That's when she first saw the photograph of Mohammad, stapled to his application. She turned to her friend Rayjean—a grad student at the University of Denver's Graduate School of International Affairs and the American Councils' coordinator for the area—and said, "It makes me uncomfortable to mention this, but he looks . . . well, he looks like one of the 9/11 hijackers."

"I know," Rayjean replied, sympathetically. "That's why these kids are so hard to place." And that made Ann want to host one even more.

She pulls another dandelion and straightens a tomato vine on its stake. The phone is ringing again.

"Ben, can you get that!" she yells toward the house.

Before the second ring ends, she's up, brushing off dirt as she runs for the kitchen door.

Mohammad Ibrahim Frotan taps his long fingers on the round window of the Boeing 727. How strange, he thinks—not glass, but as clear as glass. He presses his fingers against the plastic to leave a smudge, his print, and then looks out at the Colorado Rockies off in the distance.

They are tall mountains, as tall as the mountains of his homeland, with snow on the peaks, even in summer.

His eyes move from one mountain to the next, the bare rugged slopes and the brilliantly white snow, and he thinks about how cold that snow is. He knows exactly. He knows in ways he wishes he didn't just what it feels like at night, when you have to lie on the icy white surface because you're so tired you have to sleep and there's nowhere to escape from the chill and freezing wind, which blows right through the sunrise.

It's terrible to carry memories with you, he thinks. And the memory of snow has been with him since the day his family fled in the winter of 2002 from Bamiyan, a mountain province in Central Afghanistan. A year before, the famous giant Buddha statues that stood over his village were destroyed by Taliban explosives. He knew the giant Buddhas well, like good friends. As a boy, he and his buddies would climb a long path to the top of one of them and crawl out onto the crest of its head, nearly two hundred feet above the lush valley. When the Taliban's tanks lined up in front of the statutes and began firing, he and his dad were farming their plot around the far side of the mountain, the Mountain of Cries, and all he could hear were the blasts, one after another. It wasn't long before the Taliban turned its guns on the people of Bamiyan, and his family was forced to flee to the snowy mountains, carrying mattresses that they would roll themselves inside at night. A whole family rolled inside a mattress.

That place, the place of those moments, is so very far away now that he feels he can try to leave the memories behind. This is what he will do on this adventure. He will forget.

The man next to Ann and Ben at the receiving area at the Denver Airport can't be ignored. He clearly doesn't want to be. Ann looks over. He looks like blue-collar, gritty Colorado to her—black Harley-Davidson T-shirt, unshaven, three-days' growth. She thinks trailer park. He's talking a racist line to the girl next to him, about some issue with African Americans, a jag filled with the usual array of epithets.

Jerk. She leans over to Ben. "Can you believe that guy?" Ben doesn't want to acknowledge any of it. "Yeah, Mom, of course."

But Ann doesn't let it lie. "Watch," she tells Ben, and turns the sign she's holding so the jerk can see it. An instant later, his eyes wander to the poster board with the large black block letters: welcome mohammad ibrahim ishmael frotan.

The man's eyes meet Ann's. She smiles sweetly. Ann Petrila, small-time crusader, *in your face.* Ben rolls his eyes; his patience has all but run out. It's not just that his mom is holding a sign in an airport. She's been holding it for nearly an hour. No sighting of Mohammad. His dad—who met them here—just left to go back to work.

The crowds flow by and Ben looks intently at everyone with a complexion that matches his estimation of someone from South Asia. He's seen the photo of Mohammad, but as it is with unfamiliar features, the face is hard to conjure beyond skin color.

Ben seeks out a gate agent and then runs toward the baggage claim, vanishing into the crowd. "Ben!" Ann yells, and a minute later, he reappears, walking alongside a tall, lithe boy with very large eyes—*my, what a handsome boy*—in a YES T-shirt. The boy walks purposefully toward Ann and hugs her. "Hi, Mom," he says in a soft voice, and Ann feels like she might start to cry.

"Well, hello to you, Mohammad."

He laughs. They call him Mohammad. He's always used Ibrahim. So maybe in America he will be Mohammad, he thinks, a new person.

So an experiment of sorts begins—an experiment in cultural fusion, or collision—as the trio makes its way across an ocean of airport carpeting, across more pavement than Ibrahim has ever seen and into a rusty white car, a '96 Honda Civic.

It's as though Ibrahim has new eyes. Everything is strange, imprinting itself on him. The lady who takes their money as they leave the parking garage is black. He's seen photos of them, black people, but here she's talking, exchanging pleasantries with Ann. And there are bushes outside shaped like giant teardrops—a bush sculpture?—and then a highway overpass, with one highway going over another. He sees downtown Denver in the distance. What sort of people could build such a city? He's seen images of America, of its tall buildings, but the towers seem to rise impossibly, holding up the sky.

Ann is making conversation. How was his flight? Did he eat? She's very nice, and so is Ben, and Ibrahim— sitting in the front seat—is cheery, grinning so wide it hurts, giving short answers in his best grammatical English. They stop at a traffic light and he sees, across the intersection, a pickup truck with a young woman in the front seat sitting between two young men. He can't take his eyes off of them.

"They're everywhere," he says, under his breath. Ann turns, concerned. "What?"

"Men and women, together," he says, grimly, averting his eyes from the truck. Ben laughs from the backseat and Ann waves him off, but can't help smiling. She knows this will be an issue—she's been reading materials from the American Councils and trolling the Internet. Ibrahim, as well, has been warned. During the monthlong orientation that Naeem Muhsiny and others led in Kyrgyzstan, the students did skits to prepare for the mixing of the sexes. Ibrahim was an unenthusiastic participant, especially compared with the more sophisticated kids from Kabul or Kandahar, who make up most of the selectees. Bamiyan, after all, is a remote, rural area and home of the Hazara—descendants, it is believed, of peoples from Persia and Mongolia. Though the country's third largest ethnic group, the Hazara are second-class citizens in Afghanistan. But they pride themselves on being clear and pure in their Shiite faith. Though they were persecuted by the Taliban's religious radicals, Bamiyan's Hazara embrace a simple, traditional flavor of Muslim practice, as does Ibrahim and his family. And that practice says, quite clearly, that it is wrong for a woman to consort with men who are not members of her family.

"That's not right," he says to Ann, "that they're all so close." The girl in the truck is laughing, touching one man and then the other. And Ibrahim finds bits of Koranic passages coming into his head: the one about women "lowering their gaze and being mindful of their chastity," and another, that he thinks starts with "O Prophet! Tell thy wives and daughters that they should draw over themselves some of their outer garments" to be recognized as decent.

"You're going to be seeing a lot of that here," Ann says. "Men and women in America are together, all the time. It's just different."

He knows that, yes, and they are difficult to look at. But maybe he needs to look, he thinks. This is part of why he's here, to see things that you don't see in Afghanistan. Does it hurt him to see it if he doesn't *do it*? And why does the trio in the pickup truck seem so happy? The questions start coming fast, piling up unanswered. Ann drives them to and fro, showing Ibrahim the town and the high school—Denver's huge East High—where he will enroll in a week. She hears him whispering next to her. "Listen to what people say. Let them talk first."

It's that whisper, sotto voce, beneath wary eyes and a tight smile—cautious, reaching forward, pulling back, reaching again—that will guide the boy on this journey from the hills of Afghanistan to the heartland of America, from one end of the earth to the other. And it is by the distance traveled in the coming year that hope, in a fearful world, might well be measured. Are the divides, so clear and troubling, surmountable? Maybe, maybe not.

But first, there are gifts to give, a ceremony that occurs in the finished basement of a house in this Colorado suburb. Ibrahim begins unpacking. Michael has just arrived and Ibrahim hands him, the father of this house, a gift from his father in Bamiyan: a long black vest with stones and shiny embroidery. He hands Ann a jacket, ornate black, with gold threads woven into it, made by his mother. And for Ben, a *kofi*—with beads of many colors—which he puts on, jauntily. Ibrahim has only a small fabric bag and a backpack, so there are not many garments—a few formal shirts and slacks, four T-shirts, a *salwar kameez* (the traditional South Asian long shirt and baggy trousers), and a formal pin-striped suit, black with white stripes, which his uncle, a tailor, made for his trip to America. It's a zoot suit, patterned after something Ibrahim saw in a video.

They watch as he begins to hang his clothes. When Ibrahim turns and sees Ann and Ben, side by side, looking at him, mother and son, it makes him think of his mother—how she told him to be so careful, and to always think of her—but now he can't seem to summon her. Too much happening, too fast. Suddenly he's digging through his bag—it's in here, he knows it—and pulls out a tape, a cassette, and hands it to Ben. "Can you play this?"

Ben looks down at the cassette, back at Ibrahim, and plunges into a closet across the room, digging out shoes, winter hats, gardening gloves. He pulls himself upright with a dusty cassette player and hands it, smiling and urgent, to Ibrahim, who shoves in the tape while Ann and Ben sit awkwardly on the bed, an audience in their own home.

There's a moment of quiet crackling as the thread turns, and then the song—an ancient song about the wind crossing mountains—fills the room and wraps itself around the boy, who can't help closing his eyes and racing backward, retracing his long path, lifting his hands as his body moves. And suddenly he's back there, and here, and nowhere, dancing, in this window-less basement, the dance of Afghanistan.

Chapter 3
American Dilemmas

Consider Ibrahim's shoes. They're a rubbery synthetic. He bought them in Bamiyan. You have to look carefully to see that they cover only the tops of his feet. In the holes that have formed in their soles—and there are several in each shoe—he has packed stones. It's a trick he learned during difficult days in Afghanistan, when he had to live by himself. That was three years ago. His family had already spent three years on the run, after the Taliban slaughtered many of Bamiyan's elders, including Ibrahim's grandfather, a village leader. His mother, father, and three younger siblings—two sisters, one brother—were living hand to mouth in a nearby province. Come spring, food was scarce, the winter stores having long ago vanished. One morning, when the family awoke hungry, Ibrahim told his parents that he was returning to Bamiyan. They all knew that the Taliban spared children, meaning he might be able to get back to the family's land to plant crops. He would stay until fall and return, carrying with him the fruits of harvest. His father and mother were against it. They argued for hours, until the boy's obstinacy prevailed, and he jumped on the back of a passing truck. When he arrived home, he found that other families had had the same idea. Bamiyan was slowly filling up with teenagers. A kid town. One of Ibrahim's friends tried to run his family's bakery. Another friend sold herbal medicines. And one boy, a clever kid, taught Ibrahim how to fix his shoes. That's where he learned the trick. Come autumn, Ibrahim paid a driver to drive him and his produce—potatoes, corn, radishes, peppers—to his family. That winter, they ate.

He thinks of all this at the Major League Soccer matchup between the Colorado Rapids and Los Angeles Galaxy. Going to the game is the first thing Ibrahim does after sleeping through jetlag for a day and a half. Thirty hours of restorative sleep, then seventeen thousand screaming Coloradans in Denver's new Mile High Stadium. The place is huge, mostly empty, and vast beyond measure to Ibrahim, like a mountain range or sweeping plain. He is happy to visit it, to be on the move, with his family—with Ben and Mom and Dad. America feels enormous, and he's moving inside its valley. They all seem to be soccer fans, and they have good seats about ten rows up. In the form he filled out last year in Afghanistan, he listed soccer as one of his interests.

The teams come out and begin kicking warm-ups— a beautiful mild Saturday night. Then, as the game is about to start, a row of nearly naked women run along the side of the field. Ibrahim turns to Ann, his mouth dry.

"Cheerleaders, Ibrahim!" she shouts over the din. "They dance, and lead cheers. I know. It's a little much."

The woman—a dozen of them—almost all have yellow hair. The game starts, but he can't take his eyes off of them. They're dancing, shaking their bodies in every way imaginable. Only their breasts and behinds are covered with tiny strips of sequined cloth, and they're jumping, bouncing. There are four men with them who are clearly not their husbands or brothers, and each man lifts a woman up by putting his hand up between the girl's legs. Ibrahim wonders if he might faint. He looks down at his hands, counting fingers, and then at Ben, who has noticed—as has Mom—that he's slowly being driven to distraction.

"This is a good game," Ibrahim yells to Ben, turning away from the girls. "Yes?"

"And those cheerleaders are nice, too," Ben shouts back. "Aren't they?"

Ibrahim just nods, weakly, as he tries to control his raging, tidal reactions, and that's when he focuses on his shoes. He presses his feet against the stones, and remembers when he was alone in Bamiyan, trying to remember how his father taught him to plant seeds in rows. Then he thinks of the other children. And he gets through the quarter and then to the halftime. They go get pizza during the break at a concession stand. The pizza's tasty, and Coke is his favorite drink. As they walk back to their seats, down the long sloping aisle, he is talking to Ben about drinking Coke at the Frankfurt Airport. Then he looks up.

Dogs? Dogs are running across the field! He turns to Ben. "There are dogs!" Anything but dogs.

Like many Muslims, Ibrahim believes that the Prophet viewed dogs as vermin, believed that black dogs were the seed of Satan, and even called for dogs to be exterminated. Not surprisingly, there is nothing in the Koran's six thousand verses that frames human-canine interactions. Several pointed passages, however, appear in the Hadith, vast commentaries of Islamic interpretation that over a thousand years have hardened into cultural assessments and choices. That's why you won't find dogs as pets anywhere, for the most part, in the Arabian Peninsula, South Asia, Indonesia, or other Muslim lands. Of course, in America they are man's best friend, with 70 million as pets in the United States alone.

Right now, ten of those best friends are crisscrossing the field at Mile High and catching . . . *what?* They are Frisbee-catching dogs! People are cheering madly. The cheerleaders are now getting into the act. Ibrahim has come across the world to a place where naked woman dance with Frisbee-catching dogs. He must never tell his mother, his real mother, about this.

Driving back, Ann, trying to lighten things up, tells Ibrahim about the family dog, an old dog they got at the kennel, a mutt, which died on the Fourth of July. He was in a picture of the family that Ann sent to the American Councils' office in June to pass along to Ibrahim. He never got it.

"If I had seen a picture of your family with a dog, Mom," Ibrahim says, reluctantly, still reeling from images in his head of the leaping, jaw-snapping beasts, "I don't think I would have come."

That night, Ann thinks through the collisions of the day. She's always felt like the leader of her household, but now she also feels she's a representative of American motherhood—an American Mother bringing her Afghan "son" into the fold of humanitarian understanding. No doubt, she's got added advantages: a master's degree in social work. Her job is, after all, bridging divides. Michael, her husband, is home a bit less than usual; his work seems to increasingly dominate his time. But as the experiment called Ibrahim unfolds, she can focus her attentions on the unique conundrums at hand. Let's see what she can muster. Sitting at the computer, she trolls Islamic websites and, that night, decides to keep a journal in her night table drawer. Then she sees it. An open house tomorrow at the Abu-Bakr Mosque of the Colorado Muslim Society, advertised online for "people of all faiths."

As she stands in the beautiful ornate mosque the next day, her affection for Islam is warmed and nourished. An Imam stands and speaks to the assembled, an audience of four hundred or so—half white, she estimates, half assorted non-Caucasians. The event, planned long in advance, is now ideally timed. The crowd is large, and cameras from Denver's local affiliates are there, doing a story about Muslim outreach after the foiled British plot. Imam Ammar Amonette calls Islam "a moderate religion that has always accepted people of all faiths." It has "been around for fourteen centuries," he says, and the extremism of the past few years is "something real" but aberrant, and part of a larger fundamentalist trend in religion today. There's a break, and a buffet table of delicious fare. Ben and Ann eat chicken and lamb dishes from Afghanistan and Pakistan, hearty and spicy food accompanied by small vats of tabouleh and fresh pitas—and Ann looks over admiringly at Ibrahim, trolling the line with them, filling his plate. They are part of the solution, she feels, and getting an A for effort. And that will be an inescapable conclusion that Ben, her slightly eccentric and sensitive high-school-senior son—into Magic the Gathering, a popular fantasy card game, rather than, say, rap music and college football—will carry with him. And Ibrahim has got to be liking all this—hundreds of people celebrating Islam on a sunny Sunday. Once the plates are cleaned, they gravitate toward the mosque's president, Mohammad Noorzai, a man in his sixties, originally from Afghanistan, who is greeting visitors. He and Ibrahim are introduced and begin to speak what Ann figures is Farsi. Actually, it's Dari, the Afghan dialect of Farsi, and then Ibrahim—switching to English so others might hear—proudly tells Noorzai that he reads Arabic. The man is impressed. He gives Ibrahim a copy of the Koran. Ann is aware from her late-night trolling that most people in South Asia don't read Arabic and therefore can't read the Koran in the original. And she finds her faith in Ibrahim's abilities, and his innate strength to manage the cultural collisions and intellectual challenges ahead, rising. Noorzai talks about how Ibrahim might regularly visit the mosque, and he details upcoming events the boy might enjoy. The young man, bowing gently, holding his gift of the Koran, thanks him.

Then they are in the Honda, driving back. Ibrahim, next to Ann in the front seat, is unsettled. "I won't be going back there," he says. Ann's confused; she mentions how nice Mr. Noorzai seemed. "He's Sunni," Ibrahim says, grimly. "He's not like me. We will not speak again."

Ann exhales, silent, reassessing. *Sunni.* There you have it; an ancient blood feud inside of Islam, and the day's a total loss. She looks down at Ibrahim's shoes; he seems to be pressing his feet hard against the floor of the car. Yesterday she noticed the shoes and was appalled. Tomorrow they have to register Ibrahim for classes at East High. She turns the car toward a massive suburban mall. Today Ann Petrila is damn well buying him some American shoes.

On September 10, Vice President Cheney strolls into the Washington studios of NBC. He doesn't appear on the signature public affairs show *Meet the Press* unless there's a special need or occasion. He was on with Tim Russert a few days after 9/11, saying that the United

States would have to fight this new enemy on "the dark side." He was on in 2003, talking about the successes of the early days of the Iraq war, and then again six months later, to challenge critics who were beginning to say the Iraq campaign might turn out to be lengthier and more costly than expected. That was his last appearance.

This morning, he needs to forcefully remind America that the nation is at war—a new kind of war, the war against terrorism—and that only he, and Bush, and their party can keep the country safe.

"Let's, let's go back to the beginning here," he tells Russert, after they exchange cordialities. "Five years ago, Tim, you and I did this show, the Sunday after 9/11. And we learned a lot from 9/11. We saw, in spite of the hundreds of billions of dollars we'd spent on national security in the years up to 9/11, on that morning, 19 men with box cutters and airline tickets came into the country and killed 3,000 people. We had to take that and, and also the fact of their interest of weapons of mass destruction and recognize, at that time, it was the threat then and it's the threat today that drives much of our thinking, that the real threat is the possibility of a cell of al-Qaeda in the midst of one of our own cities with a nuclear weapon, or a biological agent. In that case, you'd be dealing—for example, if on 9/11 they'd had a nuke instead of an airplane, you'd have been looking at a casualty toll that would rival all the deaths in all the wars fought by Americans in 230 years. That's the threat we have to deal with, and that drove our thinking in the aftermath of 9/11 and does today."

After a few minutes of this, Russert picks up Cheney's then-versus-now thread: "When you were on this program, you did talk about being on the dark side, that we're going to have to get involved in intelligence and do some things with shady characters and so forth. Is that what we've done the last five years?"

Cheney seems delighted. He shifts in his chair, and bears down: "We have done everything we could think of to make the nation safe. That's our number one obligation. The, the oath that the president and I take when we're sworn in up there on Capitol Hill is always to support and defend the Constitution of the United States. And we've done everything within our power and within the Constitution to in fact pursue that objective."

Click. Usman turns off the television. "How about s*hut up!*" he shouts at the set, but not loud enough to wake up Linas and David, both of whom are still crashed after a late Saturday night.

He became a faithful *Meet the Press* viewer during his college days majoring in international relations, and has kept it up since, with his taste for knowing and forcefully debating what's going on in the world.

But something has changed. In the month since his arrest, he feels like, again, he's living events—something he hasn't felt this strongly since he stood, frozen, in the student center at Connecticut College on 9/11. A few hundred kids had gathered by dinnertime in the giant lounge in front of a large flat-screen TV. Kids were sitting in clusters, some holding each other, a few of the girls were crying. Usman, who had stayed in his room that morning, had heard by midafternoon that everyone was certain it was al Qaeda. He'd seen scattershot reports on the Internet of Muslims being attacked and Arab men being rounded up. As he walked into the large room, as a newsbreak flashed up: a Taliban official, ten feet tall on the screen, was at a podium denying responsibility on behalf of the Taliban and al Qaeda. Some kids booed. It was clear the official was lying, a proud, grinning man from South Asia in the region's traditional dress, in a long high-collared shirt, baggy pants, the *salwar kameez.*

Usman was wearing one, too; he wore it everywhere at college. Not that he would typically wear one back in Pakistan. At his private high school, the Aitchison School, they wore blue blazers and ties like little British gentlemen—the idea, in impoverished, racially homogenous Pakistan, was to be convincingly elitist. But freshman year spun him in a fresh, surprising direction, with the campus's diversity dialogue fitting neatly with his genuine distinctiveness—as one of a few dozen Asian Muslims among nineteen hundred students. Usman took on the role of a representative of Islam in a Christian society and set about, with his usual ardor, building a visible identity and authentic posture. He put on the *kameez*, kept abreast of the burning debates inside of Islam, cultivated a small patch of hair on his chin, became vice president of the Muslim Students Association and, this year, its president.

Looking at the screen, he would have traded it all for his Aitchison blazer. He felt like running, hiding. He could see in his peripheral vision that a group of kids were moving swiftly toward him. He froze—staring straight ahead—braced for the attack. This is the way of the world, he thought, the way it ends, the way *he ends*. Alarms exploded in his head—don't look over, let them land the first blow, then fight your way out of the room. A girl, a big girl, was the first one in his face, nose to nose, an inch away.

"I'm so sorry," she said, searching Usman's eyes. "This must be so terrible for you." He wasn't sure he'd heard her right. But then she was hugging him, and a boy behind her, a tall dark-haired kid, was saying, "If there's anything I can do to help you, anything, just tell me, okay?" As others stepped up, offering sympathy and asking what they could do, Usman started to breathe again. One moment, dread; the next, a sense of renewal, even salvation—a sensation so strong that he wasn't really sure if he'd ever felt at home in America until that moment.

And that's when he wanted to call his father. He knew how his father's mind worked—he was paid to be cautious, to be wary, a worrier—and he wanted to tell his dad that it was a tough day, a terrible day, but things were going to be okay.

He called a number in Baton Rouge, a number he got from his mother. Oddly enough, this was a few days into Tariq Khosa's first trip to America since his fellowship in 1987, the year he and his family lived in Seattle. He didn't have a cell phone—the number Usman called was a Louisiana State Police barracks where Tariq and thirty other senior law enforcement officials from Pakistan were housed for a two-week seminar, given by FBI trainers, on managing emergencies, both natural and manmade. Their schedule for the morning of 9/11 read, "Hijacking Plot Simulation," and the two-dozen-plus Pakistanis, in causal attire, milled about their meeting room that morning, downing coffee and donuts, and started their morning's work, talking about crisis management for hijackings.

A state trooper burst into the room and told them that the Trade Center had been hit. Everyone flowed out into the lobby and gathered before a television set, thinking it was all a simulation. "These Americans are incredibly clever," one of Tariq's colleagues said, until another one ran to an open computer, checked the Internet, and raced back in, wide-eyed. It's real. Moments later, the second plane hit, and soon the Pakistanis joined hands with the FBI agents and Louisiana state policeman, cops from across the world, comforting one another, watching every officer's worst moment. The next morning, thirty Pakistani officers gathered in a brief ceremony, standing at attention in a moment of silence and saluting their American colleagues with tears in their eyes.

It was three days later before an Amtrak delivered Tariq to New London, Connecticut, and a taxicab to Usman's dorm. "Are we okay, my boy?" he said, filling the doorway. Usman leapt into his father's arms, "Oh God, Dad, I can't believe you're here." They talked without sparing breath for hours, Tariq telling of his experiences on 9/11—of the kinships formed in Louisiana—and Usman telling of the moment in the student center, and the sensations that were racing through him. A father and son, far from home. That night, a law enforcement chief from Pakistan slept in a Connecticut dorm room with his son, in a world gone mad.

When they emerged the next morning, an amazing thing happened. Tariq Khosa—who had a day before his flight from Boston's Logan Airport to Lahore—held a revolving tutorial in the dormitory. Some of the kids had heard about what Usman's father did for a living; others heard that day. And they crowded into the dorm's lounge area and asked him questions. "Who are the Taliban?" "Are there radicals in Pakistan?" "Do terrorists usually stage many attacks in a row?"

Tariq explained it all—including the way the United States and the Saudis funded the mujahideen to challenge the Soviets, and how Pakistani intelligence officials helped build them into today's Taliban, how there were currently extremist elements in both Pakistan and Afghanistan, and how Tariq and his men chased these violent radicals into the no-man's-land between the two countries. Usman looked on with pride. The Khosas, he thought. Just look at the Khosas now.

Because, in fact, it was their world—violent, tribal, and faith-based, with its vast history and aching heart—that had now come to America.

All of this is what Usman thinks about, sitting on the couch, refusing to turn *Meet the Press* back on. He likes to think about those few days at Connecticut College every 9/11—about his fear and doubt and then his moment of renewal. But this year it's a need, a deep need. He feels himself drifting since his arrest. He's more attentive to what people see when they look at him. He's heard Cheney before, plenty of times. It never affected him much. Were Cheney's views really informed by experience—had he ever actually sat, really sat down, with a young Muslim man, or anyone like Usman? Doubtful. Now, though, Usman knew, acutely, that views are no less powerful, or consequential, for being uninformed. He was captured by such views, literally, and had he answered any one of his interrogators' foolish questions in a way that deepened their suspicion, who the hell knows where he'd be now? So he thinks of that girl wondering how he "must feel," and really wanting to know, in order to push aside the fresh recollections of how he felt in the interrogation room, the feelings of helplessness and shame, which seem now to follow him everywhere.

That night, Ann Petrila carries dishes to the sink and tells the boys there's ice cream for dessert, that they can get it themselves after they do the dishes.

This is part of Ibrahim's education, now a month into his journey. Dishes, laundry, yard work. He resisted at the start. This is not work men do in Afghanistan. Ann's response: "You're in America now; men do everything here. Not like this at your home?" Ibrahim couldn't help but laugh. "No, my sisters and mother do everything." But Ibrahim dove in, and Ann considered it an important victory.

As they all bring bowls of ice cream back to the table, she broaches the subject of the date. "Tomorrow, Ibrahim, is September eleventh—it's an important day."

Ben leaps in, much to Ann's delight, explaining what happened five years ago. "People, at school, might ask you questions tomorrow, Ibrahim. Or, even, you know, say things to you that might not be easy to hear."

Ibrahim looks at him, perplexed. "How do you mean?"

"It's a really emotional day, still, for a lot of people."

"Why?"

Ann has been detecting, over weeks, that Ibrahim's English might not be as strong as she thought. He's a master of the mechanics of grammar—understanding the placement of commas and rules applied to prepositions in ways that remain a mystery her and Ben and to most Americans—and he speaks precisely, with what sounds like a hint of British reserve. But there are gaps in his comprehension.

"Ibrahim," she says, trying to bring it home, "al Qaeda attacked America and that started the war in Afghanistan."

He shrugs, and nods, seeming to have only a vague recollection of the September 11 attacks.

Ann and Ben, both, are startled. "Why did you think people started bombing Afghanistan in the fall of 2001?"

He looks at her evenly. "People were always bombing us, for my whole life."

Ann says nothing for a moment but feels a kind of chill. My God, she thinks, how could 9/11 barely register?

Then she wants, urgently, to explain what's been happening in the wider world, the world he's in now, her world. She grabs the salt and pepper shakers and improvises a disquisition on religion—on how "your religion, Ibrahim, is facing a struggle between people who are more strict and less strict, and they're different flavors, but both the same, both spices."

Ibrahim nods. "Oh yes," he says, darkly, "I know about this. About some people who don't follow the words of Koran." Ann's not sure what to make of his tone, whether he's siding against fundamentalists or more mainstream Muslims. But she pushes on, trying to stress the invention behind so many religious distinctions and the common root of Judaism, Christianity, and Islam in the prophet Abraham. "The guy you're named after, Ibrahim."

He looks at her with a half smile, not sure if she's joking. Ben jumps in, mentioning a few boys from school whom Ibrahim has met—one Christian, one Jewish— and how they both have Abraham in their stories. Ben and then Ann tell the story of Abraham, the first Jew, and his covenant with God.

Ibrahim looks at Ann and then Ben. He clearly has never been taught any of this. "This is very interesting," he says, rising from the table, visibly agitated, taking his ice cream bowl to the sink. "Very, very interesting and amazing." Then he disappears into the basement, as Ann and Ben now turn their astonished faces to each other.

Like it or not, everyone walks in George Bush's shoes on September 11.

It's the way it works when any one person is so utterly blended into a moment in history, a date that repeats itself over and over, resonating through living memory until everyone who was alive that day is gone.

Wearing those shoes, if they happen to be yours,

is a kind of solemn obligation to the living, and to the dead.

It was, after all, Bush's day, too. Part of the American saga of this age is the improbable tale of a bully's heart breaking. Everyone saw that. He's always been a bit of a bully. Just ask his brother, Jeb, his mom, his old buddies from Yale. No one would tell you otherwise. And America detected that in him, along with the bonhomie and vengefulness, the insouciance and impulsivity. Gore, though lively in private, offered a flat public persona. That was a main reason why Bush was able to wrestle him into a virtual electoral tie; in a president, in this era of public survival through continuous storytelling, people want someone who might surprise them. Like a high-wire act with no net, Bush made it to the top mostly on pure nerve.

Which he lost on 9/11. That was visible to anyone who saw him on the tarmac making his first timorous statements and speaking uncertainly at first before the rubble at Ground Zero. This began to turn when he grabbed the bullhorn. By the time he delivered the best speech of his presidency, two weeks after the attack, he was rebuilt, a chastened bully, who wiped away tears, brushed off the dirt, and was reconstituted by vengeance dressed up as high purpose.

The moment was so cathartic for Bush it's easy to see

how it would be difficult for him to move past it. Seizing on the moment to set in motion policies such as the war against all terrorists, everywhere, or an excuse, finally, to get Saddam Hussein—efforts Bush led but had forceful company in creating—eventually caged him. It corrupted real emotion with tactical convenience.

Big anniversaries—five, ten, twenty-five—are mileposts to stop and think, to reassess the journey up to this point and consider the path to the next milepost, far ahead.

And that's what many Americans are thinking about on this 9/11. All right, five years. Where are we now, where might we end up?

Bush seems to be mindful of the subtlety of this process—a personal moment, replicated endlessly—and he treads lightly. He and the First Lady visit the World Trade Center site only on the night of the tenth, and without speeches or fanfare, they launch floral wreaths into reflecting pools in the footprints of the downed skyscrapers. With bagpipes sounding patriotic notes, they walk from the North Tower site to that of the South Tower, and after setting the wreaths adrift they return to their waiting car, which takes them to a private interfaith memorial service at nearby St. Paul's Chapel. The service and a firehouse visit close out Bush's day, and he says little to reporters other than that he's heading into the fifth anniversary with "a heavy heart."

The next morning, Bush stays clear of Ground Zero. The loved ones of those who lost their lives in the World Trade Center gather there for the yearly ceremony of remembrance. Mayor Michael Bloomberg, leading an array of political figures, gently broaches the idea of moving past 9/11. "For all Americans, this date will be forever entwined with sadness," he says in his closing remarks. "But the memory of those we lost can burn with a softening brightness."

Bush spends the morning breakfasting with New York police officers and firefighters on the Lower East Side. After this he heads to Shanksville, Pennsylvania, to mourn with the families of those who died on United Airlines Flight 93. In a cold rain, out on the field where the plane crashed, he hugs survivors and talks with them, sharing emotions for as long as they want to talk. Bush concludes the day with a trip to the Pentagon, where he lays a wreath nearby the spot where American Airlines Flight 77 breached the building.

It is understandable if Americans watching the live coverage of him moving, one among many, through the crowds of mourners, or gently titling back the wreath at the Pentagon, might wonder if this anniversary will mark a change. In the proxy relationship of a leader and a people, Bush seems, today, committed to allowing the native emotions of the day—shared national emotions, evolving steadily through the traditional process of acceptance and renewal—to guide him.

But it is not to be. At 9:00 p.m., sitting in the Oval Office for what will be his most watched speech of the year, the president uses 9/11 once again to rage against America's enemies and justify his policies, a ferocious rendering of a speech heard many times. "Nineteen men attacked us with a barbarity unequaled in our history," he opens, sitting behind his desk, looking hard at the camera. "They murdered people of all colors, creeds, and nationalities—and made war upon the entire free world. Since that day, America and her allies have taken the offensive in a war unlike any we have fought before. Today, we are safer, but we are not yet safe. On this solemn night, I've asked for some of your time to discuss the nature of the threat still before us, what we are doing to protect our nation."

And on he goes from there, saying how we've learned the nature of our enemies, that "they are evil and kill without mercy—but not without purpose. We have learned that they form a global network of extremists who are driven by a perverted vision of Islam—a totalitarian ideology that hates freedom, rejects tolerance, and despises all dissent. And we have learned that their goal is to build a radical Islamic empire where women are prisoners in their homes, men are beaten for missing prayer meetings, and terrorists have a safe haven to plan and launch attacks on America and other civilized nations. The war against this enemy is more than a military conflict. It is the decisive ideological struggle of the 21st Century, and the calling of our generation. Our nation is being tested in a way that we have not been since the start of the Cold War." After that, he speaks of "not distinguishing between terrorists and those who harbor them," like the Taliban, of how "the regime of Saddam Hussein was a clear threat," and how "just last month" al Qaeda was "foiled in a plot to blow up passenger planes headed for the United States."

Citation of the airliner plot comes at about seven minutes into the eighteen-minute speech, which is already well along in petulance, seasoned by a touch of self-defensiveness. The way his original emotions have been caged and corrupted is clear in each sentence. At this point, rating surveys show viewers switching the channel in significant numbers and turning off their sets.

Moving on its own natural arc, the country is in the process of leaving Bush—his bullying impulse fused, permanently, with satisfying vengeance—in the scattering ashes of 9/11. That's the story of this fall. The high purpose his angry words carried after the attacks, and in two elections since, is dissolving with each passing minute.

Candace Gorman wonders why it's so hard to find an Internet café in Geneva. She figured there'd be one on every corner. This is Switzerland, after all, the snow-capped peak for all things discreet and civilized and high-tech, the neutral land of chocolate, numbered bank accounts, and sex therapy.

The specific term that brought her here is *haven*. Or rather, that's what she thought of when she was having dinner with the parents of her son's best friend. They're Swiss, and their buddy, a Swiss guy, is one of the world's leading liver experts.

Candace has been reading everything she can find about the liver since she saw Mr. al-Ghizzawi, Mr. G, in July. He was yellow. When do people get yellow? That's easy—liver problems. Jaundice, hepatitis. In August, she told a hepatitis expert in Chicago what she'd observed, and he gave her an affidavit—a sort of long-distance, secondhand diagnosis—for the filing she was pulling together for the D.C. District Court. The idea was to get a court order forcing the doctors at Guantánamo at least to examine Mr. G. If a verifiable diagnosis showed he was seriously ill, she could use that to drive a habeas corpus petition. But after she saw the Chicago doctor, she was having dinner with Karen and telling her about the case—which is nothing special because it's pretty much all Candace talks about—when her son's friend's parents mentioned this doctor friend, and as soon as she said, "Switzerland," it clicked. If Candace could get the renowned Swiss liver specialist to champion Mr. G's cause, maybe she could eventually leverage that into a meeting with Swiss government officials about the possibility of asylum for Mr. G. At least, that was the plan.

And everything worked swimmingly at the start. The September filing to the district court, with affidavits from the Chicago doctor and the Swiss expert whom Candace interviewed long-distance, got results. The court ordered a doctor in Guantánamo to examine Mr. G and report back. Lo and behold, by late September there was an official medical report from Cuba. It said Mr. G had "a history of hepatitis B," that he'd also picked up tuberculosis in Guantánamo Bay in 2004, and that "his condition had stabilized."

But around that time, Candace's habeas corpus hopes vanished. On September 29, the House affirmed the Senate's version of the Military Commissions Act, which nullified any potential habeas rights of the Guantánamo detainees and reaffirmed the authority of the military tribunals to decide who was or was not an enemy combatant—a designation that couldn't be appealed in court.

Meanwhile, letters from Mr. G about his worsening medical condition were piling up on Candace's desk. Her Arabic translator, a grad student, was gone on a fellowship. But the thing is, people come through for you if you give them half a chance. That's Candace's philosophy. Because she was telling all this to Muhammad, her hairdresser at Kiza, a salon in downtown Chicago—not on Michigan Avenue, but very nice—and he said he'd translate the letters. Muhammad, who had come to America from Iraq when he was a kid and spoke perfect English, had, of course, been hearing about Mr. G since Candace took him on as a client. While he was dyeing Candace's hair, he told her that he couldn't "believe that America was treating this poor guy this way. This is not the America I came to when I was a boy, Candace, no, no." So he was delighted to translate the letters, though his Arabic was only so-so after all his years in America. He had to work about ten hours on each letter. But that's the thing about people. If you give them half a chance, they'll come through.

Which is what Candace is thinking as she searches for the Internet café—because she's just finished meeting with the Swiss liver specialist, and he's the nicest man, and brilliant. He's been making calls to people in the government for a few weeks. And, sure enough, Candace has a meeting with an official in the Swiss foreign office in a few hours.

But she's just gotten an emergency text message from her paralegal. She's received an electronic document with the military tribunal's report on Mr. G. This was something else Candace had requested of the district court. She'd petitioned the court to at least see the unclassified version of the panel's findings.

Finally, she spots one: an inviting Internet café, with enticing pastries at the counter, which she walks right by because she can't get to a computer fast enough. She's been thinking about this document, after all, since last Thanksgiving, when she took on Mr. G. It's the official record of the government's case. After nearly a year, after all the letters and their hours together, she can now get the measure of her client.

She calls it up: "The unclassified summary of evidence presented to the Tribunal by the Recorder indicated that the detainee is a Libyan citizen who has traveled extensively throughout North Africa and the Middle East and is a member of the Libyan Islamic Fight Group (LIFG), a designated foreign terrorist organization. He also possesses substantial historical and current knowledge, up to the time of his arrest, of LIFG membership and operations. The detainee visited the Khalden and Sada training camps. Afghan Intelligence Forces arrested the detainee in Konar, Afghanistan in January 2002."

Candace reads intently through the file. She knows that Sada was a camp affiliated with al Qaeda, though mostly for Afghan fighters. But Khalden had been a main training camp for al Qaeda. She isn't sure what the status is of the Libyan group. That she'll have to check. In any event, Mr. G never told her any of this, beyond his having traveled a great deal across North Africa and the Middle East. But that was when he was in his late teens or early twenties.

Then she sees something odd. A three-judge tribunal that was convened on November 23, 2004, panel #23, unanimously determined that Ghizzawi was *not* an enemy combatant. She scrolls down a dozen pages. But, then, another panel, panel #32, which convened on January 21, 2005, reviewed his case and—based on new evidence— determined that he was, in fact, an enemy combatant. Candace reads it again and reduces the screen.

After holding this man for nearly three years, what sort of evidence could they have come up with in less than two months to prompt a complete reversal? The second panel's decision was unanimous as well. She decides to petition the district court to see this new classified evidence. She figures it must be some pretty dramatic stuff. But first, a Swiss diplomat awaits. Candace feels confident the meeting will go well. She certainly knows how to be diplomatic.

She has a few minutes for one of those chocolates and a cup of very strong Swiss coffee. It's a funny thing, she thinks a moment later, as she prepares a few notes on international law and grants of asylum. It seems that the only people who *don't* come through these days—these days of suspicion and generalized fear—are government officials. Candace eats her chocolate and wonders why that is.

America is a land of miracles. Ibrahim is certain of this. He has proof, there in his hand with his name printed on it—Ibrahim Frotan, Patron—denoting that he's a member, with full privileges, of the Denver Colorado Public Library system.

There is a library a few blocks from the house. Walking distance. And he goes many days after school. It's not the books he's after. It's the DVDs. The library is filled with DVDs, and videos, too. And they're all there, his favorites: action heroes such as Jean-Claude Van Damme and Arnold Schwarzenegger, and hits from the many studios called "Bollywood," with their soft-sell romances of handsome Indian couples frolicking across sunset beaches and rolling fields. And all of them are free. Simply amazing.

Images are the one lingua franca between the world he left and the one he now inhabits; and the action heroes are like friends, filling his imagination, egging him on.

No one more than Stallone. He was the first, the one, Ibrahim says, who "begins many things for me." The seminal event occurred in 2004, a time of modest renewal in Bamiyan, after the Taliban had been driven away by U.S. forces and the Northern Alliance. A modicum of stability was settling over the village; crops were planted that spring, and kids returned to a just-repaired school. One of Ibrahim's cousins, a man in his late twenties, returned from Iran with a satchel of bootleg DVDs and a portable player. A crowd gathered around him in the town square, where Bamiyan's lone gas-powered generator rested on a tree stump. The man turned to his young cousin to make the selection, and Ibrahim chose *Rambo: First Blood III*—attracted, simply, by the word *blood*. Soon he and a hundred townsfolk were cheering. The 1988 Stallone vehicle has Rambo leaving retirement (in a Buddhist monastery, of all things) for a secret mission to Afghanistan to rescue his longtime mentor, Colonel Trautman, played by Richard Crenna. That means Rambo rides on horseback with mujahideen, scales cliffs with bare, bloody hands, and kills roughly a thousand Russians. The movie brings rapture to the town square audience. What divine magic guided Ibrahim's hand to this selection, and what extraordinary people Americans are to have a champion like Rambo, an avenger of the Afghans. That's what Ibrahim felt. He didn't think about whether Stallone was an actor, or what was real or fake. All of that was immaterial. The sensation guiding him was a powerful emotional charge that Americans were good. And as people mentioned the film to him over the subsequent months, he'd say that someday he wanted to go to America. That was the starting point.

In homage, *First Blood III* was his first rental at Denver Public in September. But there were more, selections that moved from retail versions of his old bootleg favorites to action hero classics he'd never seen—films with renderings of the America he now stood within.

Soon the one-way transaction—of a boy sitting rapt in Afghanistan, inhaling images of distant worlds— became a kind of conversation. Now he was living inside the image.

That's when, rather suddenly, the clicking started. He got the digital camera from Ann around Labor Day. It was one of several the family had, a gift that Ibrahim, thanking her, said "will be very, very important to show my time in America." He took pictures of everything he saw, images to study later, to keep forever.

This became a fixture of the glorious autumn of Ibrahim Mohammad Frotan. *Photos.* Hundreds of photos each week, downloaded onto the computer in Ann's basement, then burned onto disks for easy storage.

There were photos of the September trip to Steamboat Springs. Ben couldn't go—he had to judge a *Magic* competition in Phoenix—so it was just Mom, Dad, and Ibrahim, staying with two other couples and their teenage sons in a large rustic cabin nestled in the ranchlands along the base of the Colorado Rockies—a rugged area that prompted Ibrahim to ask, "Is this where the nomads live?" None of it made sense to him, such as why Ann and Michael would find pleasure being in a remote, rustic cabin with a wood-burning stove and no cell phone service. The house in Denver was so much nicer.

Ann and Michael were similarly perplexed by what happened the next day on Steamboat Lake. The group decided to rent a pontoon boat, and a girl who worked for the rental company was making sure the boat had gas and that the driving instructions were clear, as well as the rules of the lake. She was racing around on the dock, passing out life preservers, when Ibrahim started taking pictures of her. About thirty of them, of her breasts, mostly. She didn't seem to notice. But Ann did.

"Ibrahim . . ." she said in a stage whisper. "Pullleeeze."

He turned quickly, startled and revealed. But then she laughed, with a kind of knowing affection. And what felt illicit to him became, well, acceptable. "I'm doing research," he said, in what he instantly realized was the best joke he had ever made in English.

Ann, that night in the cabin, zinged him back. "Don't stay up too late, you know, *studying* your research."

Then they both were laughing, a moment when Ibrahim realized that he'd never had this sort of repartee with a woman—not even his mother or sisters—an exchange in which he acknowledged his own desires.

In this wash of intimacy, he felt able to bring up something he'd been wanting to talk to *her* about, something that startled him. Along with the photo essay on the breasts of the boat girl, he had taken a picture of Ann with her arm around the shoulder of her old friend John. "John is not your husband, Mom. It's wrong for you to touch him."

"Not here, Ibrahim," Ann said, sighing. "It's not wrong, here."

But as Ibrahim clicked madly to capture what his

astonished eyes saw—to flatten everything into manageable two dimensions for *study* and review—Ann was learning, too, learning to see, as best she could, through Ibrahim's eyes.

His disdain for the rustic cabin, a favorite family spot for many vacations, was clarified a week later at the Colorado History Museum, where Ann, Ben, and he whiled away a Saturday afternoon. They stopped at a floor-to-ceiling diorama of the famous Mesa Verde caves, a labyrinth of early American cave dwellings at least eight hundred years old and dug into the side a southern Colorado mountain. Ibrahim looked at it impassively. "We have those. People live in them." At a nearby diorama of the region's eighteenth-century farming techniques, where oxen pulled plows, he said, "This is like our way of farming," clearly wondering why such a thing would be in a museum.

New things, she said to herself. Stick with new things.

Of course, Ibrahim is already racing down that path by October, having downloaded nearly a thousand photos of new things, strangely new. In each picture, he holds the camera at arm's length to frame his own face and whatever's behind him: his hallway locker, each classroom, the cafeteria, the gym, the bus turnaround. It took nearly a month for him even to attach the concept of school to Denver's venerable East High. It's a maze of vast expanses, several times larger than something called "Bamiyan University," a collection of low-slung buildings in the mountains near his home. The gymnasium is an interior space surpassed only by the airports he's passed through. Mostly he just wanders from class to class and smiles. East High has a large International Club that he, two dozen other exchange students, kids from "host families," and assorted local teenagers of foreign birth are members of. That provides some structure, with a teacher in charge, a weekly meeting, and some kids who know his name, and he theirs, for hallway hellos or cafeteria table refuge.

But on a particular Monday in mid-October, Ibrahim is feeling a bit queasy, snapping, from his hip, a few passing photos from a place he hasn't yet been: in front of a class. He has to make a presentation to fellow students about Afghanistan, one of the few requirements, along with taking English, for all American Councils students.

Thank God he's long been a whiz with Photoshop—having built his skills on it and countless other bootleg versions of expensive American software he downloaded back in Bamiyan. He starts to flash up a neat, photo-adorned PowerPoint, with short, flawlessly punctuated bullet points. These are his strong suits: computers and the rules of grammar. Years on the run, not attending school, left wide gaps. Though he was at the top of his class in Bamiyan, he struggles in every subject at East.

But none of that seems to matter this morning as "My Life in Afghanistan" flashes up and, slide by slide, the kids grow quiet.

They seem fascinated by everything he shares, from stories about how "the security situation is very bad" and how it has knocked out electricity and health care, "which we need more of," to the photos of his family, which he quickly clicks through, self-conscious about how they're dressed and their modest surroundings. The questions come from every direction—from "Do you have bicycles in Afghanistan?" ("Yes, of course") to "Can you drink wine in Afghanistan?" ("No, no one can. All drinking is against Koran," a response that elicits moans from around the room.)

Afterward, a girl approaches him. She says hi in Farsi, that her name is Jasmine, and that her family is Iranian, though she was born here. Ibrahim is uncomfortable talking to girls, but when Jasmine addresses him in Farsi, he offers an enthusiastic hello back. She asks if she can tutor him in English; she, like every student, needs to complete a community service requirement to graduate. Sure, he says. He tells her in Farsi that he's been wanting to improve his English, that it's one of the main goals of his trip. Can they start today? And they do, at lunch. By then, Jasmine has checked with her father about whether Ibrahim can come to her house that weekend for dinner.

That night, Ann is pleased. She's been worrying that Ibrahim hasn't made any friends; he and Ben are friendly, but the bond doesn't seem to be deepening. She's noticed they don't talk at night while they lie in the basement in the dark, the way boys do when they're close. Maybe this will be good for Ibrahim, a new friend.

On Saturday afternoon, a black Mercedes sedan swings by to pick him up. Ibrahim is confused about the driver. His name is Reza, and he's also a Farsi-speaking Iranian, but he's not related to Jasmine. As they pull up to the house, a gleaming modern two-story mansion nearly the size of a city block, Ibrahim realizes that Reza is the chauffeur. He's seen this—Van Damme posed as one in a movie to get close to a rich and powerful guy he had to kill.

In the front hallway, he meets Jasmine's father, Nima, a friendly, middle-aged man who was born in Tehran, came here as a teenager, and started a company that makes sunglasses. He's divorced; Jasmine's mother lives somewhere far off. Just Nima, his daughter, and Reza live in this marbled mansion. In the coming hours, they float, all of them, from one room to the next, sitting on white leather furniture, sipping soft drinks, and speaking easily, warmly, in Farsi.

Ibrahim feels every muscle in his body loosen. It's like he's fallen through a trap door into something fantastic and almost familiar—a place that matches up nicely with the images of dizzying excess that so many movie-watching foreigners mistake for America, yet is warmed by the language and customs he's missed. A man is the head of this house. A beautiful daughter. A servant. It's like a Scheherazade tale, where a young man stumbles into the palace of a king and his only daughter, a princess whose affections are won by the pure-hearted wanderer. Ibrahim is so happy to express himself in Farsi that he can't stop talking. He tells them stories of Afghanistan, of how his oldest brother died fighting against the Russians when Ibrahim was a baby, how another older brother died of a mysterious disease, and how his family fled when the Taliban destroyed the Buddha statues. Ibrahim also tells of how the sun rises over the mountains to light his family's fields, of the things he misses.

They are all moved by his stories. And that night, after dinner, they shop. Jasmine expertly picks out a shirt for him, then four more creamy cotton polo shirts by a man named Ralph Lauren. Nima happily pays.

Ibrahim shows the shirts to Ann, who is sitting up trolling the Internet when he returns. Michael is away this weekend; Ben is out with friends. She holds up one of the shirts and rolls her eyes. "Very nice," she mutters, checking the label. "For this, you can buy five shirts at Target. That's where we shop, Ibrahim. *Target*." He thinks she's joking, but he's not certain, so he laughs. That's what he does when he's not sure—he laughs, to be safe.

And the next morning, he finally calls home—something he hasn't done in the past two months. He hasn't wanted to think of home, of what's happening there. Ibrahim told Ann several times that his little brother has been sick with some sort of brain tumor—that he's been seeing double for the past year. Ibrahim is worried about him, and Ann encourages him to call, telling him she's sure his "real mother" would like to hear his voice.

It is his mother who answers his father's mobile phone, and Ibrahim tells her he misses her and his father and his siblings, and that everything is going wonderfully in America.

"Is it as you thought it would be?" she asks.

"It's so big here, Mom. So much happens at once. Not the same sort of things that happen where we live. They are hard here in a different way."

He pauses, searching for common terms, as he hears her breathe. "But I am fine. That's the thing. I am doing great in school. I have many friends. And I met a family from Iran. They are Shia and very rich."

"Well, that sounds very nice, I suppose," she says.

All the while, he's been wondering why she's answering his father's phone.

"Is Dad at home?"

"No," she says. "He went to Islamabad with your brother."

He knows that means a hospital in Islamabad, a place his father has been saving money to take his brother to if there was nothing the doctors in Kabul could do to help him.

"I'm happy about this," he says, not wanting to ask more, and she does not elaborate. "Yes, I'm happy."

He tells her he loves her, and then the connection is severed. He hands the phone to Ann with a smile, goes to the basement, and lies on the bed, his eyes closed.

Continents collide inside the boy. His mother's voice comes from a place of privation, where his fingers once froze in the mountains, where two of his brothers have died and another may die soon. The urge to leave all that behind, the crushing memories, the sense of being helpless, is irresistible, encouraged by every particle of air he now breathes. This astonishing, confusing country seems to be all about confidence and capability, doing anything you set your mind to, about becoming sleek and ready, improved, the way you've wanted to be—and the *wanting* itself is important, a sleepless force. But it all feels like a game, a joke, when he hears his mother's voice, his voice—his real voice—tucked within hers.

A week later, Ibrahim finds himself in the midst of America's circus of self-invention. It's a holiday, Ann and Ben tell him at dinner, called Halloween. They try to explain it. "It's the greatest of the holidays," Ben says, "when you can become anything you want."

He goes on to tell Ibrahim he'll see skeletons and graves and monsters. "Kids dress in scary ways," Ben explains, "as a way to make fun of the things that terrify."

Ibrahim has seen skeletons, real ones. And dead bodies. And axes and long knives. The Taliban behead people with them. Why would you celebrate such things?

But in America it's Halloween. There's a party tonight, sponsored by the International Club, at the house of one of the club's members. Ibrahim has not gone to most of the club activities—he explains to Michael that he feels uncomfortable at events where girls are present—but Rayjean, the American Councils coordinator, told him he needs to learn how to socialize with the opposite sex, that it's one of the program's goals.

Ann points this out, but Ibrahim shakes his head. He's not doing this. Ben takes a subtler approach, pulling out a trunk of old costumes. The outfits in it look to Ibrahim like those from a play—from several plays. There are wigs and swords, fake noses and joke glasses. Ben puts on a strange hat, like a crazy crown, then throws it to Ibrahim.

"Just try it on." Ibrahim does. It looks funny. There are other hats, too, and then Ben pulls out the craziest thing—rubber sleeves that look just like skin covered with tattoos. He pulls one onto each arm. And then Ibrahim tries. They run to the bathroom and look in the mirror. Ibrahim has seen these before—lots of the villains in action movies have tattooed arms. Then Ben tries them again. He's laughing, making fearsome villain faces. And Ibrahim does it, too, laughing— laughing so hard he can't stop.

Ann is giddy. This is what she hoped for, moments like this. Her son and a boy from a distant land, together, learning that the world is not so vast and angry that they can't be boys, just boys, pretending and free.

There's a bustle. Ibrahim has an idea about his costume and runs into the basement. Ben already knows what his costume will be, and he starts unpacking it.

He worked at a magic store in town to earn the money for an elaborate costume of Guy Fawkes, the protagonist from his favorite movie of late, *V for Vendetta*. Ann saw the movie once with Ben—he's seen it about twenty times—and felt it was quite good, but troubling. She remembers some controversy when it came out. It takes place in England in an imagined future in which the government uses the fear of terrorism to grow itself into an Orwellian nightmare, employing the familiar tools of totalitarianism and slogans—about us versus them, patriotism, and the enemy within—right out of the Bush playbook. The hero is, in fact, a terrorist who wears the mask of Fawkes—the real-life Catholic rebel who opposed England's Protestant majority in the seventeenth century and was captured as he tried to blow up Parliament. The movie's point, Ann felt, was to show that terrorists are not some inhuman subspecies, but, potentially, anybody. Or, under the right circumstances, everybody. The movie ends with the Fawkes character bringing down the regime by actually blowing up Parliament, a spectacular and unsettling sight, as Tchaikovsky's *1812 Overture* rings forth and London is overwhelmed by millions of marchers in Fawkes masks. Ben and Ann have talked a lot about the movie—it speaks to their shared belief that fear is being used in America to crush dissent and expand government power. As Ben appears in the Fawkes costume, complete with the famous mask and a flowing cape, Ann gushes, "Ben, you look just amazing." She's proud of him. He worked hard for this money; tonight's the reward. She runs for the camera and returns, a moment later, to see Ibrahim.

Her jaw drops. He's wearing his *salwar kameez*, something he hasn't touched once since arriving. Ann thought he threw it away. On his head is a mullet wig, one of those short-on-top, ponytailed things that are a standard for the gritty, fuck-you American trailer park youth she sees as a social worker in Colorado. And on top of it all, on Ibrahim's head, is a tall, striped Uncle Sam hat.

"My God, Ibrahim," she says, looking wide-eyed at the spectacle. "I think we could call you 'The American Dilemma.'"

"Is this good, Mom?"

"More than I could explain, Ibrahim."

She snaps a picture of the boys.

Then Guy Fawkes, terrorist hero, and the American Dilemma—an original costume of a boy who carries within him the explosive fusion of Islamic fundamentalism and unquenchable individualist yearning—stroll, laughing and jostling, into the cool, dark night.

Chapter 4
One Way to Heaven

Rolf Mowatt-Larssen is already at a table at L'Enfant Plaza, the hotel just south of the Department of Energy. It's a good table, small and round, a place to hash things out at a healthy remove from the few other tables scattered around the café.

As a CIA case officer, he's been showing up early and selecting tables like this one, where someone can talk privately in a public place, for decades.

Today happens to be November 7, Election Day, ending a political season when the Bush administration has done its best to highlight and harness fear—the relentlessly cited August airline plot, the unveiling of the highest-value al Qaeda detainees in September, numerous campaign speeches by Bush in October, with the oft-repeated line "If you don't think we should be listening in on the terrorists then you ought to vote for the Democrats." But it seems not to be working as effectively, as potently, as in the other two post-9/11 elections. A kind of fear fatigue seems to have taken seed, lessening— albeit modestly—the effect of such clarion calls.

Why? Rolf shrugs. "People are finally beginning to see that fear is not a source of strength. You don't want someone reflecting back that they're as afraid as you are. You want someone who absorbs your fear, ingests it, and says, 'We have a real plan, this is where we're going and it's *going to be okay.*'"

"No one does that now," he says. But just this week, he saw something hopeful. A small thing. Bush said to his intelligence briefer, "How real is this nuclear terrorism thing? What are the terrorists really capable of? I want to break out their capability from our fear." The Office of the Director of National Intelligence summarily came to Rolf and asked him to write up a memo and prepare a PowerPoint—something the briefer could pass on to Bush.

That's fine, Rolf says, but questions like that, of capability, tend to slip into a predictable risk analysis based on inadequate information: "the probability of this, the possibility of that." He shakes his head. "That model has been very effective with most things. It's what the U.S. government tends to rely on. You gather evidence, data, and you assess the risks of what you face, and then you set up a plan, a response. Then it's a matter of execution. But that doesn't work with an existential threat like nuclear terrorism. On one hand, you have this limitless number of potential actors—young Muslim men, at least in this era—who can gather in cells that are so small, so inconspicuous in a vast world, as to be all but invisible. And on the other hand, if any one of them were to manage to pull off their shared fantasy—and it's shared by many of them—of detonating a nuclear device in, say, Washington or New York, well, the prospect is so daunting, so huge, as to be almost unfathomable.

"It's like a physics principle," he says, after a bit. "If you get really big or really small, Newton's laws don't apply anymore. The problem is that if Newton doesn't always apply, Newton's wrong." But even after the government has discovered the flaws of its old Newtonian view—which is all about force and counterforce, action and reaction—it sticks with it. Five years along, Rolf says, in a five-minute riff, we're still running around like headless chickens, chasing every threat we see. Our actions look reckless and badly aimed. The world's on edge, in tatters, and frankly "we're less safe than we were four or five years ago."

He's going grim, sepulchral. He hasn't touched his corn muffin.

"And when you compare the current conservatism and risk aversion of the U.S. intelligence community with the tough, rough characters that are now running parts of the world, well, we have a real problem—a real problem of our current capabilities that will, eventually, result in a nuclear moment here in America. I have no doubt about that, and it'll be in our lifetime."

He just sits there for a minute, taking a sip of coffee. The lobby of the hotel bustles in midmorning with diplomats and bureaucrats and tourists. It's difficult to fathom that all this could be gone in a flash—the monuments and the Mall, the White House or Congress's great dome—and we might not even know who was responsible.

But something else is afoot. Rolf, in the past few weeks, has been asking himself a simple but disruptive question: How long would it take for an undercover team to buy enough highly enriched uranium to make a nuclear weapon of significant yield—thirty-five pounds for a more sophisticated implosion device, a hundred pounds for a simpler gun-type design—and then smuggle the uranium into the United States? He keeps asking it, over and over. Six months? Four months?

For the next hour, Rolf delves into this concept; he's convinced that answering this question should be a central mission of the U.S. government. He talks about pulling together undercover teams and loosing them on the world's uranium markets. He thinks the president's question is a good starting point, denoting interest. But he's also brutally realistic as he runs through why various parts of the government are incapable of fielding such an initiative. And suddenly it seems that the "whys" (why do this) are settled, and he's just talking about the "hows" (how one might get it done). Rolf has, in fact, already chatted about logistics with someone named Rob Richer, who, over thirty years at CIA, has run Russian operations—where he and Rolf worked closely together—and been head of the Near East division and deputy director of clandestine operations. He left CIA in June 2005 for a senior job at the now-notorious private contracting firm Blackwater USA. The company was thriving at that point, with annual revenues topping a billion dollars, and recently decided to launch a private espionage firm, Total Intelligence. Rob is slated to become its CEO.

"I'm going to talk to Rob in a few days and have him write me up a proposal. So while I'm trying to push this forward inside the government, I want to also get a sense of what it might look like to outsource this."

That makes a kind of crazy sense—the sort of improvised, catch-as-catch-can pragmatism that results in billions flowing from the government to contractors who will get this or that job done, no questions asked. It's an impulse that often ends up with little accountability and a lot of people in trouble. But Richer is one of the most consequential and complex players in U.S. intelligence in the past two decades. He testifies regularly before congressional intelligence committees and advises top officials in government, who still tap him for his expertise and improvisational bravado. Richer's engagement will show what's doable—or not—in the private sector.

But Rolf is clearly enlivened to be planning operations again—something he hasn't done since leaving CIA. He bounds out of L'Enfant Plaza, with a kind of headlong eagerness, forward-leaning, if a touch reckless. He's already thinking about where the teams might find the uranium and how, after years of panicked search, he might finally replace supposition with evidence—evidence of what is, to burn off fear of what might be.

"Everywhere we're driven by fear, fear of the unknown, of what might be coming but is still invisible," he effuses, walking in great loping strides down the hill toward his office. "And by the time there'll be proof, awful proof, it'll be too late. This could really be a way to break the logical trap of trying to prove the negative, which, of course, you can never do. You send out teams. You give them a wide berth. You see what they find. Then at least you know what's real—and you can start living a real life."

It's a landmark occasion for Ibrahim, and that has nothing to do with it being Election Day. It has to with pants. Today Ibrahim will finally get the green pants. He's wanted them since coming to America. Before that, even. He saw a pair on sale in Kyrgyzstan and asked Naeem for an advance of the $125-a-month allowance each student gets for general expenses. Forget it, Naeem told him—he could find them in America. But they were hard to find. Ibrahim had been in several stores with Jasmine and Nima, and he hadn't seen them.

Ann had not been helpful here. They were at a store, a big store, and he described the pants, saying that they're like an army man wears, but casual—and she told him they were called camouflage pants and that she didn't like that kind of stuff. It evoked warfare and violence, she said. They didn't find them, Ibrahim suspects, because she didn't want to find them.

But Michael said he knew where to buy them, and today is the day he picks Ibrahim up from school and takes him to the army-navy store. The pants are perfect. Ibrahim wears them out and, in the car, he explains to Michael that he likes the pants "because you could be killed in Afghanistan for wearing them."

Michael doesn't understand, and Ibrahim explains that American soldiers, Northern Alliance fighters, and Taliban fighters all wear the pants. They are the sign that someone is a combatant. "Wearing them here," he says, "shows I am free."

Michael seems to find this interesting but changes the subject to talk about what happened the previous night, when some people with fliers and a clipboard came to the house to talk about a political movement, called a referendum, they were fighting for. Ann made sure Ibrahim came out on the porch to meet them, to "see how democracy works." When they left, Ibrahim asked, "Will they be killed by the government?"

Ann said they wouldn't, and then laughed and said that "it hasn't gotten quite that bad in America, at least not yet."

Driving around, Michael tries to explain Ann's joke, as well as the way voting works in America—that "here, the people have the power." Today, he explains, is the day Americans elect their leaders, and he and Ibrahim drive to a school near the house, where people are lined up to vote. The line is long. As they're waiting, Ibrahim sees a helicopter hovering overhead and grabs Michael's arm in a panic. "People are going to get shot!" he says, flinching.

Michael looks up. "No, God no! They're from the media, Ibrahim. The television station. They're doing a story about how long the lines are on Election Day."

Ibrahim tries to catch his breath as Michael explains it a few more times, his soothing voice as important as anything he says. But Ibrahim doesn't exhale until they're inside the building.

That Saturday, Ann has a full to-do list for the boys. Michael has been away the past few weekends, and the yard, covered with sticks and leaves, is crying out for attention.

"Okay, guys, leaf raking," she says, cleaning up the breakfast bowls and handing Ben a roll of Hefty leaf bags. "The rakes are in the garage. Honey, show Ibrahim the drill."

"I don't want to do that," Ibrahim says, standing up from the kitchen table. He's wearing the "camo" pants. He wears them every day. "It's a stupid idea. Why would you want to clean up the leaves?" He smirks, as if the whole idea seems preposterous to him.

Ann puts down the dish towel and studies him carefully. In the past two weeks or so she's thought she's detected an unwelcome obstinacy in Ibrahim's tone.

She's not sure if it's just teenage pushback, or something else.

"Well, Ibrahim," she says evenly, "we rake the leaves
so they don't pile up, then rot under the snow and kill
all the grass—is that clear enough?"

Ibrahim pauses, standing to his full six feet. He feels
like a young man of South Asia when he's with Jasmine
and especially with Nima, a person whose opinions
matter without question, a man, and a very wealthy
man at that.

"Okay, but there are many leaves on the trees. Would
it not be better to let them all fall?"

Ann's eyes widen. "No it wouldn't be *better!*"

"Ibrahim, you're not going to win this one," Ben
says, suggesting this is nothing more than teen slacker
inertia, as he pulls Ibrahim outside by the arm.

The next day, after Ibrahim sarcastically mimics
Ann's call for dinner in a mocking tone—*"Oh, it's time
for dinner, oh wonderful"*—Ann calls Rayjean. A week
later, after a few more such instances, Naeem is on the
phone from Washington.

Ann, Michael, and Ibrahim sit in the sun room, a
small enclosed porch at the front of the house arrayed
with wicker furniture. The phone, switched to its
Speaker function, rests on the ottoman, Naeem's voice
coming through.

"Okay, we're going to talk things over and find a way for everyone to get along." Ibrahim is told he has to file weekly e-mails to Naeem with a photo showing he has done what he said he would, starting with one for the following week about having a "new social encounter at school."

Ibrahim soon e-mails a photo of Jasmine and a description of a chat with her—an offering that doesn't fly. Naeem is getting back-channel briefings from Ann; she calls from her office when the kids are at school. The nighttime conversation about Jasmine bleeds into a discussion of Ibrahim's weekly forays to her house. He just went to a Denver Broncos game with the Iranians and now has nearly twenty new shirts, courtesy of Nima. "You can't accept these sorts of gifts from this other family," Naeem stresses. "Yes, but, they are my friends—they are Shia, they understand me," Ibrahim protests, and asks Naeem if they can talk in Dari, which they do, as Ann and Michael listen to the volume rise in a foreign tongue.

The next day, Naeem calls Ann at her office. "You see, Ann, there's a honeymoon phase for all the kids, and then—after they realize that the Hollywood version of America doesn't exist—the adjustment phase, where they have to deal with ways their new homes are different from the ones they left," he tells her. "This is especially true for a kid like Ibrahim." He explains that Ibrahim is, in some ways, more representative of young Afghan men—the kind that are fighting with the Northern Alliance or joining radical groups—than some of the other kids who came from more privileged homes. "He is the real Afghanistan."

"I feel honored," Ann quips. She talks to Naeem during the day but doesn't tell him everything. She's still hoping to restore the private channel of her early days with Ibrahim, their avenue for secret cross-cultural chitchats. When she finds the pile of printed materials near the computer before Thanksgiving, with photos of bare-breasted woman tucked within, she moves carefully, first asking Ben if they're his—"No, Mom, if they were mine I wouldn't leave them on the computer table"—and then handing them to Ibrahim.

"Are these yours?"

"Oh my God, Mom." He averts his eyes.

"Don't worry, Ibrahim. There's nothing wrong with this. You're a seventeen-year-old boy. But I see now why we've been running through fifty dollars' worth of ink cartridges a week."

She hands them to him. "You can have them back, just as long as you stop this printing thing."

He looks at her, down at the precious pages in his hand, and whispers, "Thank you."

"It's not about the pictures, Ibrahim. It's the most natural thing," she says softly, earnestly. Then shrugs. "It's about the ink!"

But as December arrives, she begins to realize it may be about her, and Ibrahim's unwieldy reaction to a strong, independent woman, an American woman. Naeem tries to frame it all in a long midday phone call. Over a month, he's connected the dots: Ibrahim listens only when Michael is there or when Rayjean's partner, Nick, is present.

He levels with Ann, explaining how "at eleven, an Afghan boy is told he's a man, that he must pray five times a day, that he carries the responsibility that men have. At that point, the mother can't order him to do things. The mother is mostly there to try to protect the sons, to side with them when they are disciplined by the father, who is head of all things. Her only response if he slights her, is 'I'll tell your father.' And the father will beat him. Men are in charge. I hate to say it, Ann, but he's acting a little like an Afghan man."

That night, Naeem is ready. He has practice in this realm of cross-cultural collision. He's watched his wife, from Afghanistan, move forcefully into the role of an American woman in the nine years they've been in the United States. In Dari, he tells Ibrahim he understands how difficult it can be to live in a country where men and women are equal and that, in America, the women mostly run the house. But the Prophet happens to be very clear on this: He says, "Heaven lies beneath the foot of your mother. If you keep her happy, you will go to heaven."

Ann and Michael hear Naeem offer the Koranic passage—*Beh hest zeri paye madar*—as Ibrahim gets quiet. "I will try," he softly tells Naeem in English.

On an oddly balmy morning in December, Candace Gorman arrives at the "secret facility," a nondescript office building in Washington.

She waits in the foyer to pass through three separate security checks. This is her first visit to the bowels of the government's huge secret bureaucracy. After she returned from Switzerland, she immediately petitioned the court to see the new classified evidence that prompted the second military panel to reverse the previous panel's decision on Mr. al-Ghizzawi.

She received the green light yesterday and didn't want to wait a minute. She caught a flight to Dulles this morning.

After a half hour of questions and completing forms, she enters a large nondescript room about the size of a small cafeteria, with tables scattered about. A military officer soon enters and places a file in front of her stamped classified. She looks at the word. It carries a kind of strange gravity. The designation "classified" stems, she knows, from a set of very specific laws regulating the handling of documents deemed pertinent to the national security of the United States. The statutes don't consider intent, as most laws do. They deal, oddly, with transferal and movement—lifting and carrying—as though the documents might be radioactive. It's illegal, for instance, to remove documents from a special viewing room, illegal even to remove from the viewing room any notes you've taken on a classified document, and on and on. If, perchance, a legal secretary were to become the dictator of a country—a paranoid dictator with absolute powers—these would be the kind of laws she'd write.

Candace opens the file and starts flipping. Nothing here is blacked out, unlike the redacted version she read in Switzerland. There are attached items in the back. That's where the new evidence is. She begins flipping from the front—from a report on the January hearing that recertified Ghizzawi as an unlawful enemy combatant—to the back. She reads the new evidence and thinks, this can't be right. She flips back and forth, and then back again. The "new evidence" is exactly the same as the old evidence listed in the non-classified filing: the citation of his having traveled a lot, having visited the two camps, and being a member of the Libyan Group—a group, she found, that was not even on the State Department's long list of terrorist-affiliated organizations at the time of Ghizzawi's capture. There *was* no new evidence.

"Those fucking liars," she whispers, through clenched teeth.

She checks one more time to make sure she's not missing something. No, it's clear. The first panel acquits him. A second panel is convened, given the same exact evidence, and convicts.

She slams shut the book. "This is unbelievable!" she shouts.

A lawyer sitting across the room looks up, surprised.

Candace, of course, engages him. She'd checked ahead of time about whether she could talk to any other lawyers that might be in the classified documents room. Maybe she'd bump into someone interesting.

The man approaches and tells her his name is George Clarke. Of course—*George Clarke*—she knows his name from the lawyer list serve. He handled the case of the Uighurs, a group who went to Afghanistan to escape Chinese persecution and were handed over to U.S. authorities by bounty hunters. Most were deemed not to be enemy combatants in late 2003 but remained incarcerated for years after, sometimes in shackles. In 2006, just before their habeas petition was to be heard in a U.S. court, five Uighurs were suddenly shipped off to refugee camps in Albania—the only country that would take them. The nearly twenty others are still languishing at Guantánamo.

"This is astonishing, George," she says emphatically, trying to control her volume. "They're saying evidence is new when it's not. All that happened was they took old evidence and then stamped the word *classified* on it. The only thing that's new is the stamp!"

George smiles, wanly. "Welcome to the club, Candace."

It's a Tuesday evening in mid-December at Barnes Richardson, and just the stragglers and grunts, the inefficient and the obsessive, are left. Reggie and the support staff left at five; most of the executives and analysts, by six. Usman, listed under "grunts," is settled in for a few more hours. He likes staying late. The phone doesn't ring. He can really concentrate, which he needs to do. He's been given work in the past year that is generally assigned to lawyers, often in fields such as regulatory law, or to executives with education far beyond his, such as Ph.D.s in economics or international finance. Usman's tactic has been to outwork them, staying late essentially to self-educate with open source reports and online tomes on relevant disciplines. He's currently thumbing through a deposition on a product dumping case and analysis from one of the law firms Barnes hired to handle a particularly tricky brief. He'll plow through a few analysts' reports and fire off a memo to his boss.

Then he'll hit the town. He's been doing quite a bit of that through the fall—quite a bit, in fact, since the day of his arrest. When he got home that July night, he wanted to shut himself in to decompress. He didn't want to talk to anyone. But his man-about-town friend Zarar, head of the Pakistani networking club, was on the phone and relentless. Usman had to come out and see his buddies—"We're all worried about you, Usman"—so he went to Café Citron, in Dupont Circle, and ordered a Grey Goose on the rocks. Usman, hoisting his tumbler, called it a "Muslim Martini." That night, everyone worked hard to draw humor from the arrest. And from that point forward, Usman came to appreciate this above all else: people, especially coworkers and non-Muslims, who joked about this his woe, showing that they didn't take it seriously. Night after night, they razzed him. It was a mishap. A farce. An outrage! The idea that anyone would see Usman Khosa as some religious radical, bent on the destruction of the apostate Unites States, is patently absurd. Who could be more

American than Usman! And then they'd seal it with another round.

At around 8:00 p.m., the calls from friends start coming in. Usman pushes aside an analyst's report as his BlackBerry rings.

It's a friend of his, Pervez, a trade association executive, who has two friends visiting from Pakistan.

"You really know the party scene, Usman," Pervez says, "and my friends want to party."

Usman tells him to swing by his office and, a few minutes later, Pervez is outside in his Mercedes with his two friends in the backseat.

After brief greetings, one of the friends states his case: "Where do we find a club, a great club?"

"Well, the clubs are kind of quiet on a Tuesday night," Usman says. "Why don't we just go to a bar."

That draws a frown. "Where do we get, you know, the best girls? Where do you go for that, for the really great girls?"

Usman nods. He's seen this before. Guys from Pakistan come for a few days and want to do everything they can't do in South Asia. It's tiresome, and born, he feels, of the way Islam can repress and twist basic natural desires. But he understands. "Let me guess, you guys want to go to a strip club, right?"

They both nod, enthusiastically. First, though, they

want to check into their hotel, and Pervez weaves the Mercedes through downtown Washington to the Hilton Hotel on Fourteenth Street. One of the Pakistani guys goes to check in as Usman and Pervez chat with the other one about life these days in Pakistan and, specifically, Lahore, where he lives. Both guys are in their late twenties, working in family businesses, unmarried, moderately religious, and otherwise unremarkable. After a moment, the one chatting with Usman and Pervez stops the small talk for an important logistical note: "Usman, hey, before I go to the strip club I want to eat, but I only want to eat halal." Usman shoots him a look of bemused impatience. "Let me get this straight. First a properly blessed Islamic halal meal, *then* the strip club." The guy nods, oblivious to any irony.

His friend returns from the front desk and says through the window of the idling car that they have to get the bags out and have them taken to the room. Everyone jumps out and helps pull the heavy suitcases from the trunk. And, yes, the guy who just checked in says he also wants to eat halal—not easy to find at eight thirty on a Tuesday in Washington—before going to the strip club.

Usman looks at them both. "You both want to go to a strip club, and you want to see some naked people—as many as possible—and you want to eat halal?"

The fact that neither of them sees any contradiction in this dual request strikes Usman as precisely what's wrong with Islam—what he more and more considers the religion's utter lack of self-awareness about how little real relevance its ancient, rigid rules have in the modern world.

"This is messed-up, man," he says to them both, throwing up his hands. "That's why I don't like Islam. It's why since coming to America—intellectually, at least—I've become a bit of an atheist!" He impatiently grabs one of the suitcases and heads for the hotel lobby as one of the Pakistani guys follows him, pulling the other bag behind him.

Pervez gets in the car with the other guy from Lahore, who now seems troubled, pensive.

After a moment, he speaks. "Usman used to be Muslim and he's not now," he says seriously to Pervez. "You know, if we kill him, we can go to heaven."

Pervez looks at him in disbelief. "What? No, no killing," he stutters. "He's just saying this in an academic way. Seriously, he hasn't converted out of the faith formally or anything."

The friend is unimpressed. "People usually don't. What's very clear is that the Prophet says that any believer who becomes a nonbeliever has committed a sin against Islam and he who slays such a nonbeliever will be granted everlasting life. It's one way to heaven."

Pervez is wide-eyed. He went to school in Pakistan with one of these guys, but he doesn't know them all that well. "Look, you're way off here. Usman is a good guy and, deep down, still a good Muslim."

The friend shrugs. "And heaven's a good place to be, too."

A moment later, they are all back in the car. Usman tells them of a restaurant in downtown Washington where they can still get dinner—good food and halal—and of a strip club not far from where he lives. But he's already eaten, and wants to change out of his suit, so they drop him off at his apartment. He'll meet them later.

As he gets out of the car, Pervez, looking oddly at him, says, "It's okay if you have stuff to do and can't make it."

"No, no problem," Usman says, and an hour later he walks into a club just south of Dupont Circle. He surveys the room and spots the Pakistani trio. When he gets to the table, Pervez looks at him like he's seeing a ghost. "Usman," he says, "look, come here, I have to ask you something about work."

"Work?"

Pervez pulls him away and leads him to a quiet area near the bathrooms.

"Usman, look, you've got to get out of here." He describes what happened in the car, how one of his friends is sure that killing Usman is his ticket to heaven.

Usman looks at Pervez in horror. "How about thou shall not kill? What? That's not in our religion? This is unbelievable! Just make sure these lunatics don't come to my house, okay? If they're on their way, you call me—immediately!"

Usman busts out of the club, races home, and locks his door. In a long sleepless night, grimly picking through memories and books on his shelves, he thinks of his grandfather and his father—their codes for living, their hopes for him—and tries to make sense of what it means to be a Khosa, in America.

Ann can't stop being herself. And neither can Ibrahim. And it comes around to the pants. She, Ben, and Ibrahim travel to Ann's parents' house in Indiana for Christmas. Ibrahim's been wearing nothing but the camouflage pants for more than a month, but he's not— repeat *not*—wearing them for Christmas. She tells him this when they're packing, and makes sure both boys have some nice clothes. Then she sees Ibrahim waltz into her parents' dining room wearing his camos.

"This is not acceptable, Ibrahim. You have to change them."

He pauses, looking hard at her. He loves these pants. They are the pants he can't wear in Afghanistan—he's told Ann this—without being in mortal danger. But in America, he is free. Wearing them gives him a sense of freedom; they are his affront to fear. This Christmas is not even his holiday!

She doesn't budge. Neither does he, until he breaks, pounds up the stairs, and slams the door.

A week later, Ann hears from a teacher that Ibrahim's reducing his class load to three classes: English, weightlifting, and art.

Ben has been accepted early decision to Knox College, a liberal arts school in Illinois. He's beset with senioritis and has become impatient with Ibrahim, who's going to Jasmine's every chance he gets. He now has dozens of polo shirts. Ann and Michael feel the Ibrahim experiment has started to envelop their lives. Both agree that it's turned sour. And they don't want Ben's last months at home to be colored by global conflict at the kitchen table. They decide it's time to call Naeem again.

Ibrahim can tell by the flatness in Ann's voice—"We need to talk to Naeem about the best place for you"—that it's all collapsing. He starts swiftly building his defenses, telling teachers at school that Ann has made him eat pork and drink wine and has kept him from calling home. Ann gets a call at her office from Brian Doyle, the teacher who oversees foreign students.

"What's going on, Ann? You won't believe what Ibrahim is saying."

When Ann hears, she blows. "This school is our community. This is our reputation."

Ibrahim doesn't take the bus after school with Ben that day. He's unaccounted for. Michael leaves work and runs through the empty halls of East, eventually finding Ibrahim and bringing him home.

Ben has slipped off to a friend's house, clear from the blast area.

And in the kitchen, opposing ions collide, best intentions exploding, affection turning to anger.

"Ibrahim, how could you be saying those things around the school?" Ann shouts.

"You've done all these things, and I have proof," Ibrahim counters.

Ann is startled by the rage running through her. "You've completely lost our trust. You've betrayed our family. We wouldn't go to Afghanistan and do this to you!"

She's met with equal and opposing ferocity. "You are a pork-eating, wine-drinking woman!"

"That's it. You can't stay here anymore," she shouts. "It is *over!*"

The word hits him like a fist, like the finality of death.

He begins to beg and bow, like a slave begging for mercy, for his life. "Please, oh God. I'm so sorry. Please, Mom. Please, please."

Ann is not sure what she's seeing, if this is some awful vestige of medieval penitence, a groveling plea from the lost centuries of darkness, of brutality—and of course it is. *It's what he came here to escape, to my house.* And now the hardness of the world, of something vast, beyond either of them, gathers velocity and hits her in the chest. "God, I'm so sorry," she gulps, tears streaming down her face.

The next morning, Ann and Ibrahim are in the car. The plan is simple, arranged in late-night calls to Naeem. She'll bring Ibrahim to the University of Denver and leave him for the day with Nick, who works in admissions at the Graduate School of International Affairs. Ann will leave Nick her car, as he and Rayjean have only a scooter. Ibrahim will spend the night at their house, and the next morning they'll drive him to the airport—destination, as yet unknown.

As she drives, silently, she thinks again and again of their screaming match in the kitchen, of how she said she wouldn't go to Afghanistan and act this way toward him, as if they were peers, equals. Whatever Naeem said about Ibrahim asserting his Afghan manhood, that was wrong. "I'm sorry about the way things worked out," she says, breaking long minutes of silence.

"Yes, me, too," Ibrahim says, in a whisper. "I am a child. You are an adult. I have made mistakes." He pauses. "But you have as well."

She nods, not saying anything for a moment as the car weaves forward. "We love you as a son, Ibrahim. We do, and we hope, all of us, to see you again."

He looks out the window at the passing landscape. "I will think often of these days."

The large American Tourister suitcase—a gift they gave Ibrahim for Christmas—is on the seat behind him, filled with Ralph Lauren shirts, a *salwar kameez*, disks with thousands of image files, and printed photographs of smiling naked women.

At Nick's office, Ann says goodbye. Ibrahim looks ahead, silent. He doesn't turn as she walks away.

Sitting in her office half an hour later, Ann is not sure if she's ever felt much worse.

Her IM flashes. It's Nick. "Hi. Ibrahim says I'm supposed to take him to the Iranians' house. I told him I can't leave the office. He wants to know if you could come over and drive him there."

Ann shakes her head. *Unbelievable!* She types deliberately, slow and firm: "NO CAN DO," and punches Send.

Across campus, Ibrahim Mohammad Frotan is locked into survival mode. Here he has hard-earned skills.

He finds a moment when Nick is busy, commandeers an office line, and calls Reza, the chauffeur: "Come get me, quick."

By late afternoon, he's on a stool at the granite island in his favorite kitchen, drinking a Coke. Jasmine gets home from school. Nima, from work. Together, the Shiites scheme. Nima tries to figure out a plan, feeling a sense of injustice about how Ibrahim has been treated. Ibrahim goes lightly on his host family. He speaks, instead, of his dreams, of a life in America. Nima is not certain of the immigration laws or Ibrahim's legal status in the United States, but he says he'd be willing, someday, to support Ibrahim in his education. Ibrahim hugs him. They all eat a triumphal dinner, celebrating their common bonds, and the future.

Then the phone rings. Jasmine answers and hands the receiver to Ibrahim.

"What exactly do you think you're doing?" It's Naeem. He's tracked Ibrahim down. Ann, of course, has Nima's phone number.

Ibrahim's mind races. He says nothing for a minute.

"I'm not going back," he says, finally, in Dari. "My Iranian family will support me. I've decided to stay in America." He's surprised he's able to speak the whole thought. That final sentence, he thinks, must have been forming inside him for years.

He knows Naeem can be fierce—he still remembers how he faced down the gate agents in Frankfurt—but now he hears an unfamiliar voice, vaguely threatening.

"I'll only say this once, so listen carefully. You are here on a visa from the American Councils for International Education. We are responsible for you. And *we* will be deciding the next step in your trip. If you take any path other than doing exactly what I say, Ibrahim, you will be in the United States illegally. That means against the law."

The word *law* stops Ibrahim cold. He's speechless, in free fall.

Naeem closes it. "Nick will be over in a few minutes to take you to Rayjean's and his apartment. Tomorrow morning you'll be on a plane."

"Where will I go?"

"There's a place in Alabama where you'll stay for a week or two while we look for somewhere, anywhere, you can go for the second half of the year." He pauses. "Of course, you know, it might be difficult to find a placement."

What Naeem knows is that the American Councils' Washington office is recommending he go back to Afghanistan—an ignominy only three previous students have suffered. He doesn't tell this to the boy.

But Ibrahim, his survivalist antennae on highest alert, suspects it. The breath seems to escape from his body.

"So are we clear, Ibrahim?"

"Yes," he murmurs. "Yes, I'll do what you say. I'll go with Nick."

For reasons unfathomable, Ibrahim then thinks of the stone shoes. A memory flashes by of the day Ann made him throw them away—how the great green garbage can shuttered as they hit bottom—a thud, easily recalled, that tells him life will be that much harder in Afghanistan, knowing all he now knows and already misses.

ACT II

The Armageddon Test

A very modern man walks the streets of Bukhara, a bustling commercial and cultural center along the Silk Road.

The year is 998. Head in the clouds, his sandals slap across dusty streets of filthy beggars and plumed merchants, vast bazaars with spices, fruits, fresh-cut lamb, and embroidered cloths alongside outdoor sewers, compounds for royal families, rows of mud huts, and the minarets of mosques, tall and ornate, in the early flowering of Islam. Through the heart of the city, a capital of the Persian Empire north of the Iranian plateau, moves a vivid parade of traders, dancing their loop from Rome to Peking to Baghdad, hawking strange products and customs and ideas, a catalytic compound that makes this city a laboratory of the human experiment.

The young man, Avicenna, has the sort of mind that

graces humankind only infrequently, augmented, in this case, by advantages of birth—his father, a wealthy politician from a nearby town in Afghanistan—and by tutors, who shaped his intellect with the best learning to be found in Persia, which includes the Greek philosophers. He's already read Aristotle's Metaphysics a dozen times, fighting to understand every word, especially the peripatetic Greek's concepts of empiricism and the power of reason.

But it's not until he buys a pamphlet from a street vendor on this sunny day in Bukhara—a commentary on Aristotle's work by al-Farabi, a predecessor of Avicenna's and a Platonist—that his world begins to turn. He pays three dirhams for the thin book, reads it breathlessly, and finally comprehends the power of Aristotle's concept of the mind as a blank slate—"a tablet that bears no writing"—and of "being as understanding." Handing money to the beggars crowding around him, he runs to the mosque to thank God.

It's a lightning strike that starts a blaze. The fire rages inside of Avicenna, who begins writing treatises to match his forebear, 150 treatises alone on various areas of reason and philosophy, some of which revive, rethink, and restore Aristotle.

For the past three hundred years, the Greek philosophers have been banned by the Church, which saw, clearly, that their ideas on how people might seek discernible, concrete truths threatened a vast power structure built on faith, now called the Holy Roman Empire. Plato was demeaned as godless. Aristotle, forgotten. Metaphysics wouldn't even be translated into Latin for another two hundred years, and then only as part of the Church's strategic plan to co-opt and crush the "rationalist nuisance."

So it ends up being young Avicenna—a marketplace philosopher—who carries the Hellenistic ideal forward with ferocity, inventing and expanding many of the ideas of empiricism by applying them to a rigorous, fearless ethic of experimentation. He turns out to be the bridge, the link, the interpreter. His book The Canon of Medicine introduces rigorous experimentation and clinical trials to the study of medicine, pioneers the concepts of contagious disease and quarantine, and lays the foundation for modern-day pharmacology.

He races forward from there, asserting, in a mountain of scholarly volumes, that men can improve themselves through the power of reason and aspire to "universal truths."

This intellectual explosion makes Avicenna famous and places him alongside the other genius of Bukhara: Muhammad al-Bukhari, a religious scholar who died ninety years before Avicenna's birth. Islam, at this moment, is in a formative period, struggling to consolidate the vast reach won by both inspiration and force at its founding. Two centuries along, the faith of Muhammad hangs like an intricate veil from Cairo to Babylon and beyond: a religion still searching for institutional wholeness, a set of lessons to live by.

The answer springs from the fierce, ostentatious al-Bukhari, a Muslim polymath graced with extraordinary recall. Today we'd call it a photographic memory. Back then, it seemed like a miracle. Al-Bukhari traveled through the varied lands of Islam inhaling the myriad recollections of the life and teachings of Muhammad. The Koran, the 114-chapter recitation that Muslims believe God spoke through the Prophet, provides stunning sweeps of inspiration but very little in terms of prescription or parameter, rules for the conduct of human affairs. That's, of course, where religions find their lasting utility. On that score, al-Bukhari is a one-man solution. After eighteen years of travel, a city-to-city tour interviewing scholars and displaying storied feats of recall before crowded rooms, he returns home to Bukhara. There he writes one of history's most consequential texts, Sahih al-Bukhari, the central collection of Hadith, or "narratives," drawn from the life and words of Muhammad.

It contains 2,602 accepted and tested traditions,

which al-Bukhari arranges in chapters, or books—the Book of Jihad, the Book of Life—to provide for a complete system of jurisprudence that lays the foundation for Islamic law and sets the rules for human conduct. Other Hadith would be written, but al-Bukhari's is considered by Sunnis to be second in authority only to the Koran, and the strictures it articulates—covering thousands of activities from birth to death, prayers, holidays, greetings, menstruation—would stand firm from its ninth-century drafting onward.

And so, springing from Bukhara's chaotic, bustling streets, the teachings of al-Bukhari and then those of Avicenna manage to march abreast, through what would later be called the golden age of Islam, a time when the countries beneath the flag of Sharia pioneer everything from lending libraries to environmental science, degree-granting universities to public hospitals. The competing camps of faith and reason created fierce open debate, encouraged as part of the notion of Itjihad, the technical term of Islamic law that refers to the process of arriving at a legal decision through an interpretation of religious texts and commentaries. The view, propounded by Avicenna and his adherents, was that religion and philosophy were separate but compatible—one, God's gift to man; the other, man's effort to attain perfect knowledge. Al-Bukhari saw perfection in God's word as passed through Muhammad, but was driven by his voracious intellect constantly to consider the urges and uplift of mankind—that poor, noble collection of dust—in order to distill divine inspiration into standards of conduct. Two men, two rule books, both respected and both flourishing.

Scholars will endlessly debate why such delicate equilibria are rare and difficult to preserve, even when, as in this case of Persia, the people are prospering through the spirited competition of temporal and divine.

But one thing is certain: disequilibrium is often instigated by the so-called will to power, a sleepless drive in the human personality to control others, to force them to do what one wants, or not do what one opposes.

In this case, a few decades after Avicenna's death, a book called The Incoherence of the Philosophers, published by a Persian scholar named al-Ghazali, started to tip the era's delicate balance. It accused empiricists and adherents of the rational search, especially Avicenna, of heresy. It alleged that they denied the true power of God and His intermediate angels over cause and effect. It attacked Greek philosophers such as Aristotle for being corruptors of Islam. The book was wildly influential, especially with the help of scores of Imams, who'd come to realize that their authority was being diluted by Itjihad and its independent interpretations of *how or whether religious principles underlay Sharia's laws.*

Over the next three centuries, the gates of Itjihad—of discourse and inquiry, review and reappraisal—began to close, and Islam's golden age started its slow fade. Al-Bukhari's Hadith, with its rules for living the just life, held firm, forming a cornerstone of Sunni Islam, of Sharia law, supporting the axis of divinity and authority that would shape many nations and that currently forms the core of Islamic fundamentalism.

Avicenna's ideas, meanwhile, would whither in the Islamic world and blow westward, to nourish the writings of thinkers ranging from Thomas Aquinas to the Jewish philosopher Maimonides—who favored the Muslim thinker over Aristotle—to countless doctors, physicists, and scientific researchers, who relied on his rigorous rules of experimentation. By the time Avicenna, Aristotle, and their co-philosophers were inspiring John Locke in the seventeenth century, the Christian Church's millennium of dominance—its iron clasp of power and faith—was fast giving way to the Age of Reason, to the Enlightenment, when people would begin to place their faith in the social contract, in the power of the individual, in man, writ large, as the measure of all things, and in earthly sacraments such as inalienable rights and informed consent. Religion didn't vanish, of course, but found its match, its irrepressible competitor. A balance was struck that holds, in large measure, to this day, in which humankind's tireless search for the knowable has remade much of the world. Artifacts of that transformation would include Avicenna's dog-eared tenth-century copy of Aristotle's Metaphysics, written in an early version of Arabic, the three dirhams he paid for al-Farabi's commentary, and any coins he handed out to the beggars.

And those sandals, because we all, in a way, walk Bukhara's dusty streets. The battles fought there between reason and faith, between God's gift of revelation and man's search for knowledge, continue to rage in every corner of the earth. As does the will to power—searching, tirelessly, for leverage and the chance to reap all that can be harvested.

Life-shaping institutions tend to start with an idea or revelation, sparked by an Avicenna or an al-Bukhari. But—from the Holy Roman Church to the keepers of the Holy Places of Mecca and Medina to modern-day governments—they so often become mostly about the preservation and expansion of authority, something they relinquish only with the greatest reluctance.

Here we briefly leave the struggles of Usman and Ibrahim—two young Muslim men who walk, each day, the fault line between faith and reason—to look at the latest manifestations of the will to power. The improbable recent twist is that man's search to understand the world—the search for the scientific knowledge to unlock the secrets of the known universe—has handed the adherents of revelatory certainty the destructive capability to reverse power shifts that have been under way for five hundred years. The drama, at this moment, involves the most powerful nation on earth and its struggle to rediscover its original, transforming principles. The battle over those oaths—to basic integrity in the public place, to accountability to the governed, and to controlling power's raging tactical urges—is visible in what unfolds at present and in a reprise of how America arrived at this impasse.

Chapter 1
People to People

The year 2007 starts with a few lessons in the tension between ends and means, beginning with the death of Gerald Ford.

The thirty-eighth president is remembered most for a single decision: when he put the country's interest ahead of his own political fortunes by pardoning Richard Nixon. Over the ensuing decades, as that decision was widely judged a sound one, Ford was heralded for recognizing that the goal of wise governance is not always reelection. He realized that the means of his daily stewardship—making ethical decisions even if they flouted partisan calculations—were what mattered most. He knew the pardon meant he'd lose in 1976. He was right.

Several members of Ford's senior staff and cabinet serve as his pallbearers—including Vice President

Cheney, Federal Reserve chairman Alan Greenspan, former Treasury secretary Paul O'Neill, and Donald Rumsfeld, who resigned as defense secretary in mid-December in the wake of the president's repudiation in the midterm elections.

That setback, and the Republican Party's loss of both chambers of Congress, was indisputably linked to the interplay of this period's two grand initiatives: the struggle against terrorism and the war in Iraq. The former—which may prove the more historically important and long lasting—has been conducted largely in the shadows, a man-on-man struggle driven by intelligence that has granted leaders a great deal of political and creative license. The latter, a more traditional conflict that involved armies clashing to secure territory and then soldiers engaging with civilians, was not providing any such political conveniences. Four years along, with three thousand U.S. combat deaths and more than twenty thousand injured soldiers, not to mention an estimated two hundred thousand Iraqi casualties, it has unfolded in stark sunlight for everyone in America and abroad to assess. No doubt, much of Election Day's harsh appraisal flowed from displeasure about the by-any-means-necessary mission to oust Saddam Hussein and the confused, ill-considered efforts to bring stability to Iraq.

Inside the National Cathedral on this landmark occasion, the burying of a president, is the entire living leadership of post–World War II America. Looking at the familiar faces—the former presidents and their famous deputies, leaders of Congress, present across all the decades of tumult—now solemn and aging, prompts reflection about what abides and what is lost with the passage of time.

This calculus is dense with fresh variables, including the ugly execution three days earlier of a man who was a nemesis of many in the cathedral. A cell phone video of Saddam Hussein, looking fit and surprisingly serene in his final moments, has been steadily circling the globe. In it, he's being bullied by Shiites chanting "Moktada, Moktada," in honor of their clerical strongman, Moktada al-Sadr, as the rope is slipped around his neck. The scene looks like a lynching, sectarian style— a far cry from the swift justice expected and hoped for from a great, civilized power. That was clearly the goal, the noble end, set at the first National Security Council meeting of the Bush presidency in January 2001, when the ways to oust Saddam Hussein were discussed in detail, as were justifications involving WMD. Six years hence, the question floating inside the cathedral's majestic nave alongside eulogies about Ford's selflessness is how political expediency—the means that power employs to augment and expand itself—may have overwhelmed deeper national interests.

Two days later, on January 4, the Democrats take control of Congress, making California congresswoman Nancy Pelosi the first woman ever to sit in the House Speaker's chair and elevating Nevada's senator Harry Reid to the office of Majority Leader. The day marks a return to divided government, the state of play for twenty-eight of the past thirty-eight years and a way that voters have long hedged their bets, knowing intuitively that neither side has a monopoly on right answers. Because so many of the self-correcting processes of democracy—such as congressional oversight, judicial review, and a robust press—have been hampered in the years since 9/11, many Americans seemed to use their vote to carry forward that reliable principle, a favorite of the Founding Fathers, that power is best expressed when it is shared. Voters demanded, in essence, that some governing authority be passed to Democrats. On this morning, that officially occurs.

Meanwhile, readers are snapping up the morning's *Wall Street Journal* to witness another passage, this one conceptual. There, on the editorial page, the longtime billboard of conservatism in America, is a column by four heavyweight graybeards: Henry Kissinger, secretary of state from 1973 to 1977 and adviser to presidents ever since; George Shultz, a secretary of state for both Reagan and Bush; William Perry, a defense secretary under Clinton; and former Georgia senator Sam Nunn, a conservative Democrat who long chaired the Armed Services Committee and has led the charge on nuclear security issues since 1992.

They are calling for an end to all nuclear weapons.

"Nuclear weapons today present tremendous dangers, but also an historic opportunity. U.S. leadership will be required to take the world to the next stage—to a solid consensus for reversing reliance on nuclear weapons globally as a vital contribution to preventing their proliferation into potentially dangerous hands, and ultimately ending them as a threat to the world."

That's the way it begins, but, in a sense, the column is really an end—a final collapse of the idea that the nuclear weapon could ever be controlled or, at day's end, kept out of the hands of terrorists.

Each man experienced his own particular evolution on the issue, but all four—two Republicans, two Democrats—experienced a kind of shared disenchantment with the cold war doctrines they'd all helped execute. The balancing of the vast arsenals of the United States and Soviet Union, and each nation's defensive compacts with allied nations, long provided order and coherence, enforced by the all-trumping possibility of global destruction. There were plenty of heated, disastrous conflicts during the post–WWII era, from Americans in Korea to Russians in Afghanistan, but there was never a head-on collision of the world's two great armies and arsenals. That was an indisputable victory.

By the time Ronald Reagan and Mikhail Gorbachev were meeting in Reykjavik in 1986, anxieties had eased a bit, fitfully but steadily across decades—from backyard bomb shelters to Nixon and Kissinger's détente policy to friendly hockey games—even while Reagan embraced "evil empire" rhetoric and plans for the "Star Wars" missile defense shield. That may have been mostly posturing, the actor's craft, because the seventy-six-year-old president shocked his advisers by treating Gorbachev as a peer, a friend, and asking if he was ready to get rid of these monstrous weapons. The resulting agreements reduced both countries' arsenals by six thousand warheads and set goals to go even further.

And that's what the four men, and a few dozen other officials, got together in October 2006 to reminisce about: the day, twenty years before, when Reagan, in his dotage, helped all the "hawks" feel like peacemakers. They met at Stanford to talk of this, their save-the-world moment, but soon the conversation turned to how the world had moved on and slipped away from best efforts at self-control.

The Nuclear Non-Proliferation Treaty, or NPT, the centerpiece of the movement since the 1970s to halt the spread of nuclear weapons, was in tatters. The nuclear club had grown to nine members, including volatile Pakistan and North Korea, which had recently detonated a plutonium bomb. Iran had begun enriching uranium in a kind of affront to the idea that any nation should have its sovereignty impinged by unprovable concerns and hypotheticals, especially those of a "hypocritical power" like the United States.

The charge was annoyingly hard to dismiss. The United States, and other members of the club's old guard, preached tirelessly about stopping the spread of the nightmare weapons, all the while enjoying the strategic leverage they provided. This structural flaw, a hairline crack in the foundation of disarmament efforts, was widening under stress—the stress of technology's relentless, borderless spread that was turning "have not" nations, one after another, into "why not?" nations.

By the fall of 2006, a crop of them were hitting maturity overnight, like rebellious teenagers, the graybeards felt, responding with a shrug to a parent's final desperate paternalism: "*Why not?* Because I said so, that's why not!" The Turks wanted a bomb now. So did the Egyptians and the Saudis, claiming that they'd need to match the Iranians, and fast. A dozen countries from Burma to Bahrain were cranking up nuclear energy programs that would, if allowed to flourish, leave them with materials that could be reprocessed into bombs or stolen by terrorists. The Sudanese were starting a program, too, and Africa was emerging as a competitive market for illegal uranium.

What those at Stanford could no longer avoid were the deeper dilemmas of moral authority. Kissinger and the others, men who believed in the realpolitik of balancing interests between nations, had never been much on the concept of "moral authority"—it was difficult to quantify in the equations of power. Now they acutely felt its absence.

Many were old enough to mistily recall victory in "the just war," World War II, and knew what it felt like to be admired. But how do you get that back? You evolve, they figured, like Reagan did; you hand over something you have, something you've gained fair and square, to support the common good. Is that weakness or strength—an avowal of the limits of force or the chance to grab a prize that only force can grant? What, in fact, they grasped in their long walks at Stanford, wrinkled men crunching across fallen leaves, was a very old lesson: that moral action—humble and honest—is the tribute that power must at some point pay to reason. Without ever admitting that their lives' work of managing force to make the world safe stood in shambles, they expressed contrition—a central feature of moral action—in the lexicon of geopolitics.

"Nuclear weapons," they wrote in their cosigned column on January 4, "were essential to maintaining international security during the Cold War because they were a means of deterrence. The end of the Cold War made the doctrine of mutual Soviet-American deterrence obsolete. Deterrence continues to be a relevant consideration for many states with regard to threats from other states. But reliance on nuclear weapons for this purpose is becoming increasingly hazardous and decreasingly effective.

"North Korea's recent nuclear test and Iran's refusal to stop its program to enrich uranium—potentially to weapons grade—highlight the fact that the world is now on the precipice of a new and dangerous nuclear era. Most alarmingly, the likelihood that non-state terrorists will get their hands on nuclear weaponry is increasing. In today's war waged on world order by terrorists, nuclear weapons are the ultimate means of mass devastation. And non-state terrorist groups with nuclear weapons are conceptually outside the bounds of a deterrent strategy and present difficult new security challenges."

In a more declaratory tone, it might read: *We, fading guardians of the world order, proclaim that nuclear weapons are a type of power that all nations must now relinquish, starting with America.*

But could disparate peoples ever come to such consensus, such shared purpose? And who would have the moral authority to lead such a mission?

A task left for younger men, younger women.

"I love their phrase—'conceptually outside the bounds of a deterrent strategy,'" Rolf says over the phone on January 5, a day after the *Journal*'s column appeared. "Someone who believes it's a messianic duty to explode a nuclear weapon in New York is definitely outside of the bounds of deterrence. Here, we're talking about incentives. Being a martyr *is* the goal."

A few minutes later, he's at L'Enfant Plaza fussing with his BlackBerry, firing notes to his scheduler. The column has set official Washington on fire. News stories are being written about it. Pundits are opining, on and on.

He sees this as for the good—it foists the issue of terrorists with nukes into in-boxes all over town, bumps it right on top for the New Year. He's busily booking appointments that'll stretch forward two months, a round of briefings for top officials in the intelligence community, the NSC, the Vice President's Office, about what's happening in the "markets," where he's sure weapons-grade nuclear materials are on the block. He'll badger them in his urgent whispers, stitch together a bit of the known into a tapestry of the supposed, vividly render the probable, making it feel real enough to touch.

This, after all, is his specialty, his improbable passion. Rolf is a madman—charmed, wild-eyed, intense and distractible, affable and terrifying. He's broadly versed, was once recruited to teach philosophy at his alma mater, West Point, speaks four languages, including fluent Russian, is a skilled pianist and a former paratrooper—a tall guy, fifty-two, and solidly built at six feet two, with a thick mop of prematurely whitened hair, a quick smile, big laugh. But he's sometimes grim, prone to deep funks, and, like anyone who's succeeded in the clandestine world, adept at working in both sunlight and shadow. Throughout his career he's been the lone voice who's been right enough times that the near-misses, the hair-on-fire rides and panic-button frenzies, have been tolerated. People have Rolf stories, some used as example and encouragement; some kept safe, on ice, for ready inspiration when a chance comes to strangle him.

Yet in many ways, he was constructed across decades for this very moment. He's perfectly suited for the schizophrenia that comes from chasing ghosts.

He's actually done it before. It's all there, in Russia: the Rolf dilemma. In 1989, as a young agent in CIA's huge Moscow station, he became convinced that there were moles inside the U.S. government wrecking havoc. People said he was crazy; what was clear was that he was driving everyone else crazy. They sent him packing, shipping him off to run the CIA station in Athens.

Then, two years later, they spotted something strange. Evidence, maybe, that there was a mole, that Rolf had been right. The CIA boss who engineered Rolf's exile was sent to Greece to apologize personally and persuade Rolf to return to Moscow, fast. Didn't take much. Rolf raced back to Moscow and dove into the counterintelligence mission that eventually helped nab Aldrich Ames.

But then it all turns, just one more notch. He started seeing moles everywhere. By 1996, Rob Richer ran "Russia House," Langley's large headquarters for all things Russian, and Rolf was his deputy. Together, they hatched a disaster. They became convinced there were as many as thirty Russian spies inside CIA. The Russians artfully gamed the situation, using the arrest of Ames to plant suspicions that there were many others: a blond man with a mustache, a tall dark guy who liked hats, a charming lady they'd recently met. Were they ghosts? Were they real?

Recriminations raced through Moscow and Washington; and, in fact, another mole or two were caught. But a toxin of suspicion had entered CIA's bloodstream. Hundreds of agents in the mid-'90s were put through polygraph tests, an imperfect measure of guilt when testing people who lie and posture for a living. As part of procedure, scores were referred for further probing to CIA's archenemy, the FBI, creating an investigatory logjam. Many agents, who felt they'd spent their lives in a noble cause, grew tired of being under scrutiny. They resented it and left the agency for easier lives. The polygraph mania was stopped in 1999. Even Rolf and Rob were relieved, though they continued to quietly assert that their suspicions were far from unfounded. Then, in the summer of 2001, the most damaging spy of them all—a man who was responsible for security breaches costing billions and the death of as many as a dozen U.S. moles and sources—was arrested. His name was Robert Hanssen, an FBI agent. Go figure. Right suspicion, wrong agency.

A few months later, planes hit the Trade Center, and Rolf, in a similar dynamic, began advising the Bush administration on terrorists seeking WMD, a subject of unparalleled gravity.

There were, again, startling clues and widespread doubts. Even though it was clear that bin Laden and Zawahiri wanted their own nuclear weapon—they'd met with Pakistani nuclear scientists a few weeks before 9/11 to talk through bomb-building logistics—it was unclear how far al Qaeda had proceeded toward its goal. Rolf and his CIA team ran down some blind alleys, probing knowns and unknowns as best they could, whipsawing between shaky intelligence and profound fears in the months and years after 9/11. At day's end, the hits—the few they picked up about terrorists and WMD—were often the secret fuel driving policies, disclosures that inhabited Bush's nightmares and, in some cases, compelled executive action.

But never enough action, never enough even to make a dent in the endemic and growing problems of what all the most knowledgeable players—and now, publicly, the graybeards—call a "new and dangerous nuclear era."

"Fine," Rolf says, drinking coffee at L'Enfant's lobby café. "Kissinger and the others are calling for global disarmament. That's admirable. But what do we do in the meantime, which could be a very, very long time?"—a time when "the current process of government is systemically unable to solve our problems.

"I always believed after 9/11—and I'm still wrestling

with this—that we can get the system to work. I know now, we can't."

This is a statement of finality, like a divorce, or the start of a new life, both an end and a beginning. Which is why he's pushing forward on two tracks.

One is inside the system: the endless meetings, the sit-downs with a largely declawed CIA now acting mostly in support of "war-fighting" efforts in Iraq, and the briefings to senior officials who are often dispirited and fatigued by problems with no ready solutions.

The other track is outside the system, and it could unleash havoc.

That's what he's very quietly talking about with his old partner Rob Richer, who's busily drawing up plans for "the teams." The teams need to look and act just like terrorist cells, or groups a terrorist network such as al Qaeda would hire to locate and buy enriched uranium. Rolf is feeling, more and more, that Rob is the most natural person to head this sort of operation. And it's not just because Rolf feels that CIA is currently incapable of staging such a mission. The teams, he feels, will succeed only if they're what corporations often call a "skunk works" or "tiger team" solution—denoting a special unit gathering together the best talent to solve a particular problem, but outside the "conventional structure, which," Rolf says, "can't get it done." And it's not, he elaborates, just because the U.S. government would want to deny knowledge of the operation—though it might.

No, it's because the teams would actually be running a test—a real, incorruptible test—on both the U.S. government and the world at large. And in its current state, the U.S. government can't be trusted to run a test on itself.

"It's better to go to an outside entity, for it to have a simple, arm's-length quality," he says, "all based on a clear goal." Here's your mission. Here's your money. Here are some basic rules we'd like you to follow. Now go buy the enriched uranium, wherever the sellers may be, and smuggle it back into the United States. Then call in.

This would be Rolf's vision of "mission accomplished"—terrifying, crushing, in a way. But real, finally real, burning off the ghosts and the what-ifs. At the center of it all would be a kind of hard, irrefutable truth that the U.S. government and its minions can't deny or table or spin or bury: the fact that it can be done—that uranium can be bought in sufficient quantity to build a bomb and brought into the United States—and now it *has* been done. Done in an operation, an experiment, that was not shaped by the self-defensive, politicized urges of those in power—elected leaders whom you, citizen in America or a foreign country, may not believe and whom Rolf himself is lately having trouble trusting. "This test would be a shock to the system," he says, and "it needs to be very public, without a lot of fingerprints on it." Because, Rolf asserts, it's something that everyone, here and abroad, should know. Because then, all together, across all the man-made borders, "we'll do what's needed to save ourselves."

The Guantánamo Bay ferry chugs across a small harbor separating the wider world from the newly refurbished detention facility and bumps up to the prison dock. The sun glints off the water, a gentle southerly wind warms the air.

Candace is met by an escort, a smart young JAG officer in his mid-thirties, a handsome guy like the ones on the TV show. His name is Kevin, and a few minutes along, she's sure he's the most civil person she's met on her four trips. She's concerned that Ghizzawi was moved two months ago, to Camp Six, where about one hundred inmates pass their days in solitary confinement, a kind of supermax at Guantánamo, built by Halliburton for the most hardened, "high value" detainees and discipline cases. Her client is neither.

"It doesn't make sense," she tells Kevin, looking to confide in him. He says he understands her concern and that it's more about logistics, finding places for all the prisoners. He tells her a few unusually candid things—such as how chaotic it sometimes gets at the prison—and then he tries to turn it into a joke. "Just think of it like your guy has his own apartment," he says, with a smile.

"Don't you say that to me!" she snaps.

She's not to be trifled with, not now, not after seeing the actual case against her client and that he was, in essence, acquitted and then reconvicted on the same thin evidence. She's starting to feel that this entire process is set up and that she's representing an innocent man—a rarity, really, for a lawyer—that Mr. al-Ghizzawi is, in fact, a baker with very bad luck.

A baker who doesn't want to see her. That's what the guard says once she gets to the building with the newly built interview rooms. "No, detainee Six-five-four said he doesn't want to see you. I'm sorry, ma'am."

Candace is befuddled. She turns to Kevin. "Procedures allow me to write him a note. And, yes, he can read English." Kevin reluctantly nods his assent to the guard, while Candace scribbles on the back of a sheet of her legal pad: "I know Mr. al-Ghizzawi that it's been difficult being put in Camp Six. I know your health has deteriorated. But if you could meet with me for five minutes, it could be helpful for your situation."

A minute later, she's escorted into the freezing interview room. Ghizzawi is lying in the corner, his arms wrapped around his legs, shivering.

"You have to get up into the chair," the guard tells him. "Either you get into the chair or your lawyer has to leave." Ghizzawi doesn't move. The guard, some American kid from the Midwest, it sounds like, just looks at him for a minute and then drops his guard tone. "Well, maybe we can get you a cushion or something. I'll be back." He leaves Candace with her client.

She's not sure what to say.

"I didn't want you to see me like this, Mrs. Gorman." His teeth are chattering. "I'm so embarrassed about being like I am. In the orange suit." The orange jumpsuit is for disciplinary cases. He's now judging himself, harshly, by his captors' rules.

"Why don't you just tell me what happened," she says, sitting down.

He explains how he was taking a shower two days ago and had a handful of toilet paper in his pocket. It's against rules to have anything in your pocket when you go to the shower room. A guard saw it and said it was a violation. He was stripped and put into an orange suit. When that happens, they take away your thermal shirt as part of the punishment—the thermal shirts that were granted three months ago after two dozen lawyers filed complaints. There are no blankets in Camp Six; there were, but they were taken away. Now Ghizzawi's ashamed and freezing.

And that's making his illnesses—his untreated hepatitis B and tuberculosis—even worse. Though it had been confirmed by camp doctors that he's suffering from both, he has yet to be treated. Candace has continued filing petitions demanding medical treatment, but she keeps getting no response. Ghizzawi lies on the concrete floor, his leg chained to a cemented ring, holding his side, whispering, "I am so sorry . . . so sorry."

The room is tiny, only about six by six, and much of it taken up by the table between them. After a few minutes, he manages to sit up. And she talks to him a bit. He says he's vomiting two or three times a day, has constant pain in his abdomen, and often soils his clothes, which he tries to wash in the toilet.

She asks if any interrogators are seeing him anymore. "No," he says, "there's nothing else for me to tell them."

Looking at him, now so diminished and despairing, she thinks about the many stages he must have passed through, only the last few of which she's witnessed. She tries to imagine him in the early days, how he might have looked and acted when they arrested him, and during the initial interrogations, the first humiliations. What was lost when they stripped him—he and twenty others—and made them act like dogs, hoods over their heads, on the tarmac in Afghanistan; or in the early days at Guantánamo, when they beat him with chains, threatened him with death and rape, bound him for hours in excruciating stress positions, and humiliated him with body-cavity searches in front of laughing, picture-taking guards? What does this do to a man? What does he shed and what abides? An hour passes, and another.

He mentions that he's received a letter from his brother. He has an older brother in Libya. Candace has been trying to find him, but she can't. Somehow the brother found Ghizzawi—possibly through the International Committee of the Red Cross, which tries to visit each detainee every two months or so. He shows Candace the letter with his brother's phone number at the top.

Candace thinks this must have cheered him. Not so. He's fearful that it's a trick, that the letter is not authentic. He says she should call the number and ask two things: whether the man's daughter died, and what year it happened.

"If he hesitates, hang up," he says darkly, his humor,

his touch with irony, having given way to enveloping suspicions. The two months in solitary confinement, chilled to the bone, have taken their toll.

A few more months like this, and he'll be a shell. For the first time in decades, she's lost faith in the law providing any remedies, at least anytime soon. It seems, more and more, that this entire structure is built to keep the law at bay, at arm's length. But that thought is like a seed that can find no purchase, even in Candace's skeptical sensibilities. How, for instance, could a nice young lawyer like Kevin—even with his comments about the studio apartment—be part of a system *designed* to wrongly imprison people? The United States doesn't build systems like that—not intentionally. At least, she can't imagine that it would.

She says she'll call his brother as soon as she gets to a phone, and takes notes on his symptoms, the conditions in which he lives, his general state of mind. Maybe, she figures, she'll be able to use what she's seen to impel to action civil rights groups, or foreign governments, or maybe just readers of her blog.

That night, sitting with a half-dozen other lawyers in the hotel on the edge of the base, she finds that the discussion of the legal strategies, of tactical moves and petitions to file, has been reduced to the hypothetical.

In frustration, she blurts out, "Maybe I'll just go to the NEX (the navy store) and buy my guy a goddamn blanket. I'd get arrested, but it'd be worth it—at least I'd be doing something!"

With the passage of the Military Commissions Act in the fall—one of the last major acts of the Republican Congress—the only thing the detainees can challenge is their status as enemy combatants, a status determined by a military tribunal system with almost no legal obligations other than what it sets for itself. The MCA's dismissal of the writ of habeas corpus leaves the lawyers virtually powerless to help their clients.

Except in the most basic way, the way a visitor—any visitor—might lift the spirits of a prisoner. The next morning, Ghizzawi seems, in fact, to be modestly revived. He asks Candace about her family, her daughters and son, her husband, who, she told him two visits back, is a sociologist. What kind of sociologist? Then he talks a bit about his family, about how his wife and he worked opening the bakery and what the weather must be like this time of year in Afghanistan.

After a while, he asks, almost in passing, "Does anyone in America care about us down here anymore?"

It's the type of loaded question that would usually prompt a swift Gorman comeback—something to inspire, to stoke a fire: "Of course they care, and that's just what we'll use!" This time her resignation carries the moment. She sighs, telling him that most people are focused on the death, last week, of a has-been model named Anna Nicole Smith and the question of who will be declared the biological father of her baby, an infant girl, who stands to collect millions. As she goes through a lengthy explanation of who's who—feeling embarrassed that she knows as much as she does about the whole frivolous mess—he smiles, slyly, like he used to. "Do *you* know who the father is?"

Delighted, Candace snaps to. "Don Rumsfeld's sudden resignation is very suspicious, but they're still waiting for the lab results."

And they laugh, like nothing. Like two friends, chatting about nothing at all, and that feels suddenly like revenge.

As spring approaches, the government of the world's most powerful nation is running furiously in place.

Three great campaigns are under way. The military effort in Iraq will enter its fifth year in March; and the war in Afghanistan and the broader struggle against the emerging force of global terrorism will both enter their seventh years in the fall.

Inside the country's vast federal colossus—an organism that consumes more than $2.7 trillion a year—it is difficult to ask about original principles, or define basic terms.

Justice, in particular, is a term with a history as interesting as that of any word. It appears, often repeatedly, in virtually every significant human text. In the Hebrew Bible, it is mentioned twice in the phrase "justice, justice, this you must pursue," to stress that justice be one's guide for both the ends and the means. In the Christian Bible, justice is often framed by compassion, acts of charity and grace to the less fortunate. In the Koran, the very concept of justice is one of God's bequests to man.

It is the Greek philosophers, though, who may have brought the greatest effort to unpacking the word, to shaping it. That foundation of Western philosophy, *The Republic*, by Plato, deals almost entirely with simply trying to arrive at a basic definition of justice, offering many, including honesty, fairness, humility, and respect for the law. But Plato, whose goal was to build his imagined Republic on a foundation of justice, ends his treatise by folding many concepts beneath the mighty word, and defining it, by the book's end, as something like "integrity"—which he describes as the integration of many parts of the human personality. Essentially, if one is integrated—fitted together around an array of basic, rather common virtues—he or she will be ready to act justly.

It's a very long way between Plato, who all but invents many of these conceptual cornerstones, and the most powerful nation ever to be built on them. But defining justice as integrity—the integration of essential qualities—is oddly applicable across time.

The government of the United States is, at this moment, wildly disintegrated—uncertain about which original principles should define action, much less how they might fit into a coherent whole. This drift and internal conflict, Plato might say, makes it almost impossible to act justly. The country's three great campaigns work each day at cross-purposes, the left hand undermining the right: the war in Iraq fuels global terrorist recruitment and siphons essential resources from the Afghanistan effort; anxious countries in the region meanwhile busily build or buy arsenals—some with WMD—to slake insecurity about their neighbors and unrest within their borders; the underlying insecurity and unrest, in turn, are born of and compounded by all the above.

Government contractors and lobbyists, throughout, profit handsomely from the confusion, while the dozens of message departments across the government—in maybe the only example of coherent purpose—send out the good news, day after day, that all is well. The major change of late is the arrival of a Democratic majority in Congress. They've already launched investigations that the executive branch is working furiously to derail, while day after day they issue the counter-message that all is lost. Also untrue.

Rolf sits in his office at sunrise on March 9. Look at his desk—it's all there. If there's a WMD attack on the United States using some sort of nuclear device, all of it, the whole desktop, will be the first thing subpoenaed by Congress. On one side are reports about three dozen significant attempts to steal and sell fissile material, mostly uranium, since 1994. The pile is up to a standing man's shoulder. What the pages show is the sustained effort, year after year, of hustlers and traders, common criminals and sometimes respectable men, often government officials who grew desperate in ugly circumstances. The incidents are mostly in Russia and the former Soviet states, so the circumstances usually have to do with the dissolution of that particular military empire—an empire that collapsed not so much because it faced a competing military force in the United States, but because it couldn't unleash and nourish the economic ingenuity of its people. The country, with its massive nuclear arsenal, collapsed because people were desperate, and it's those same needy people who often pop up in Rolf's reports, digging up some radioactive nugget they've buried in their backyard and looking for a way to cash it in for a car, or a couple years with enough to eat. Necessity and desperation birthing resourcefulness, and then a sit-down with some passing reprobate who says he knows a buyer. The reports show a market that, like all markets, is evolving, and becoming more professional, proving itself surprisingly resilient.

Rolf has been involved in what may be the most harrowing of the incidents—integrally involved. In 2003, a package was delivered to him at CIA. A man was picked up crossing from Russia to Georgia with 170 grams of uranium; he said his customer was "a Muslim man." A Georgian official friendly with CIA went to a courier service in Tbilisi. Uranium through the mail. Rolf's CIA team got the package and tested it: 93 percent enriched. That's extremely high. Sixty-five percent enriched is enough for a detonation. This was the gold-standard uranium, the kind the Russians used for their ICBM. CIA sent a few folks over to Georgia and soon traced the uranium to Novosibirk, the Siberian hub for nuclear production facilities. The Russians denied this publicly—said they knew nothing about the uranium's origins—but privately told CIA that they were on the case. In fact, Putin told that to Bush in one of their calls. And a year or so later, Putin told Bush in another call that the Russians had wrapped up the conspiracy and plugged the uranium leak. Everyone felt satisfied. But then, in February 2006, there was another seizure in Georgia. Same thing; different guy. That presented a host of unsettling conclusions. Three years after this uranium-trafficking network was detected, and despite several assurances from Putin, it *still* had not been shut down. *Three years*, and the smuggling—from Novosibirsk—was still going on.

In January 2007, when news of the second Georgia incident leaked out, Rolf briefed members of Congress. What he didn't tell them—except those with the highest security clearance—was the deeply classified finding, known to only a few people in the intelligence community, that the incidents were from the very same network. That left only a few people who could ask the question: What might have gotten through? You can smuggle a lot of uranium in three years.

While the pile of incident reports grows, another pile—on the other side of the desk—rises to match it. It's a pile of printed reports with declarative statements, agenda items, resource requests, and worthy goals: *briefing materials*. What are they worth? Hard to say. Many of the proposals deal with relations between states, their bureaucracies, their police, intelligence agencies, and, at day's end, leaders—a multilateral ideal that has a poor track record apart from those rare moments of demonstrable crisis. But a big part of Rolf's job is to crank out these sharp-edged reports. He's good at it—a product of all that liberal arts learning—and even better at presenting them.

Which is what he's been doing nonstop since the start of the year. A few weeks after the graybeards' column, news of the Georgian smuggling incident came out, care of a frustrated Georgian government anxious about Russia's growing influence in its breakaway region of South Ossetia. Georgia's interior minister disclosed the details of the case during a trip to Washington, where he hoped there'd be enough concern about nuclear materials and porous borders to prompt international action on the South Ossetia problem. No one moved to resolve the separatist dilemma, but the news did put nuclear smuggling front and center. Rolf was called upon to brief various committees in Congress, though only those with the highest security clearance were told that the 2003 and 2006 incidents involved the same network.

The interagency group he's been heading for a year—a monthly colloquy between top officials from all corners of the intelligence community called the IND (Improvised Nuclear Device) Steering Group— sent out a classified report to the White House and all sixteen intelligence agencies called "Markets: Goals and Methods." It breaks down strategies for infiltrating the markets along traditional divides: *unilateral* (just the United States trying to figure out what's happening in the markets with basic intelligence work); *joint* (which is doing the same thing in partnership with other countries); *private* (intelligence contractors, such as Blackwater); and *other.*

All that, in governmentese, makes up the background materials for Rolf's face-to-face collisions, meetings like the one he just had with Stephen Hadley, the president's national security adviser. Rolf set a trap, a PowerPoint saying that it was clear that stopping nuclear terrorism was "not the nation's first priority." Hadley, taken aback, responded, "No, it *is* our first priority," which triggered Rolf's dam break of data and despair, all spoken with utmost respectfulness—an exegesis on how nothing, really, is being done and how our "leading indicators" suggest that terrorists are probably out buying materials, at this very moment. Hadley then wanted more, much more, including reports to bring directly to the president. This dance was repeated in a meeting with CIA director Mike Hayden, though Hayden took it a bit more personally, because at the center of the current assessment is an understanding that CIA, with its diminished clandestine service and creeping check-the-box military rigidity, has little to offer.

But Hayden's new boss, Mike McConnell, had a different response. McConnell, a former head of the NSA, became the second director of national intelligence in February, replacing John Negroponte, a career diplomat with no intelligence experience, who all but raced out the door and back to the familiar realm of the State Department. Rolf was only halfway into his presentation when McConnell, fully engaged, cut to the chase. "Well, what are we doing about it?" he asked. Rolf looked at the clock. There wasn't enough time to run through all the ins and outs, so he sketched a brief overview and promised to send along more comprehensive materials. McConnell cut in again, saying that, yes, he got it—he didn't need more charts and memos to see how insubstantial the current approach was. "Doesn't look like enough to me," he said. "Looks like we don't even have a plan!"

That statement, coming from the top intelligence official in the nation, encouraged and affirmed a flurry of effort. Nothing broad or structural, no rethinking of fundamental principles, but well-intentioned forward progress.

One plan that Rolf and CIA consider involves subtly engaging Russian organized crime bosses, including the more public ones, several of whom have achieved glamorous Al Capone–like status among Russians. Most notable among them is Semion Mogilevich, who now runs Russia's largest criminal syndicate out of his offices in the UK and Israel—a massive multibillion-dollar operation that includes arms dealing, drug trafficking, prostitution, and money laundering. Mogilevich, who holds a degree in economics and is sometimes called "The Brainy Don," has been hiring carloads of ex-CIA agents to shore up his operations as he moves into more respectable ventures. A foot in shadow, a foot in light. In this, Rolf sees opportunity. He and a few CIA chiefs sketch outlines of how to approach Mogilevich, how to impart the message that a black market nuclear theft would be "bad for his businesses." No quid pro quos that would grant the thug anything—of course not. But any efforts he could muster to identify smuggling networks would help challenge the complacency of Russian officialdom and would be "duly noted" by U.S. officials.

A more promising initiative, subtler, slower, but more conceptually interesting, is to engage gray area organizations—including certain Muslim charities, humanitarian groups, refugee and rights groups—to get a sense of their needs and to express, in confidence, the specific fears of the United States. These groups have mostly been targets—targets of largely unsuccessful prosecutions and harassment by the U.S. government. This hasn't yielded much, aside from ill will toward the United States and a heightened status for the organizations, making them more invested than ever in cultivating an anti-American posture.

Dealing with such organizations slams against ramparts erected after 9/11 by the U.S. government and, most notably, by Vice President Cheney. In umpteen instances, he's halted such initiatives for committing the sin of "rewarding bad behavior." But global affairs since 2003 have made it eminently clear that the rest of the world is so mightily rewarding anti-American behavior that no one has much incentive to curry favor with the United States. In fact, the opposite tack has been the more profitable, as everyone from Putin to Ahmadinejad can attest.

With the sun now burning off a morning mist, Rolf stops by L'Enfant Plaza. He has an hour before a 10:00 a.m. briefing he's due to give five department chiefs at the State Department. He talks about these gray area initiatives and whether he'll mention them in today's mega-brief, but he knows that'll put him in the crosshairs of the Vice President's Office, the OVP, which he has also been briefing regularly. He's not sure he can afford to stand openly in opposition to the OVP, and what he calls their "behavioral fixations." Many top officials in the government are beginning to feel this "don't reward bad behavior" idea has been an unmitigated disaster— one that flies in the face of principles that have guided U.S. foreign policy for years, old commonsense notions of engaging the enemy whenever possible.

But there's more to it than that. Something doesn't fit. Clearly, a major reason the United States invaded Iraq was to compel "rogue states" who support terrorism—or, for that matter, those who support extremism or regional interests that undermine U.S. authority— to change their behavior. The central idea was that Saddam Hussein, largely defanged but still petulant, was an easy mark, an ideal candidate to be a "demonstration model" to alter the behavior of other regimes that might consider opposing the United States.

But even as Rolf goes through this—an accepted analysis of Iraq in the upper reaches of government by 2007—he's shaking his head.

"There's more, here. Something under the surface," he says, in a moment of unwelcome doubt after months of motivational briefings, the next one scheduled an hour from now. "Some things still just don't add up."

This sends him backward to a place he's visited many times: the engagement with the Iranians in 2003. Like any intelligence agent who sees something irregular, something incongruous, Rolf goes back, reconstructing events.

Iran has long been the real power in the Middle East, the flashpoint of the Islamic revival that swept the Muslim world after the rise of Ayatollah Khomeini in 1979. A country of 70 million with technological prowess and oil wealth, and the home of the once mighty Persian Empire, Iran owns one of the planet's proudest histories. Of late, they've been enriching uranium in opposition to the world community and, according to a few recent statements by the U.S. government, supporting both anti-U.S. insurgents in Iraq and the resurgent Taliban.

But in the spring of 2003, things were very different with Iran. Everything was different. A few hundred miles from Tehran, U.S. troops were racing across Iraq in the first heady days of the invasion. Baghdad had fallen in a lightning-quick military victory over Saddam's troops that proved easier than anyone could have imagined, and then Saddam's statue went down in one of those perfect symbolic images.

Meanwhile, a parallel track was invisibly moving along: ongoing conversations in late 2002 and early 2003 between two key arms of al Qaeda—one in Iran, one in Saudi Arabia—which were being closely monitored by the NSA. Al Qaeda's Shura Council, a kind of management group for the organization, had fled to Iran in late 2001, during the last days before Tora Bora fell. The group included Sayf al-Adl, one of bin Laden's senior deputies, al Qaeda's operational commander, and a true "number three" beneath Zawahiri; and Abdel al-Aziz al-Masri, an Egyptian scientist who ran al Qaeda's nuclear program. The Iranians assured the British that these parties were all under a kind of loose house arrest. But the Shura Council also happened to be communicating regularly with their al Qaeda colleagues in Saudi Arabia, who were soliciting advice about the purchase of Russian "suitcase nukes." From a luxurious villa in Iran, al Qaeda's Sayf al-Adl assured Abu Bakr, the group's Saudi chief, that no price was too high to pay—that al Qaeda leadership would get whatever money was needed for the purchase—but that he should be careful about being conned. Al Qaeda had been stung at least once, to the tune of $1.5 million, by scam artists claiming to have uranium. This time things would be different. Al-Adl assured the Saudis that they would have access to the Pakistani scientists now working with al Qaeda to verify the authenticity of any nuclear material before purchase.

This was, obviously, of utmost interest to the United States. But even more interesting was what happened when the British—having preserved their long-standing contacts with Tehran—passed along to the Iranians, at the Americans' behest, a description of the NSA's intercepts. The Iranians expressed shock, and displayed what seemed like genuine concern. They told the Brits they wanted to help, and admitted that they'd neglected the exiled Shura Council, which was operating freely inside their borders. But that could change. They were experts, after all, in their own region, with cross-border contacts as deep and dense as those the United States had with Canada, or Britain with Ireland. They were ready to start subtly working al Qaeda's exiled managers and maybe even use them to get to bin Laden, now hiding somewhere on the Pakistani-Afghan border.

The Brits were delighted, as were senior intelligence officials throughout the U.S. government. Iran was the prize catch. Secretly sitting with them for a few tough hands of geo-poker and a late-night chat about shared needs had the potential to positively alter every conversation in the region. The United States hadn't meaningfully engaged Iran since the Shah's fall in 1979. That could finally change.

What prompted the shift? Iraq. That was the analytical consensus. The United States had 150,000 of its troops next door to Iran. This did more than send a message. It altered the Iranians' way of thinking about their national interests.

Soon the British started to work on where and when a few high-level meetings might take place—to get the dialogue started—while CIA's teams for both WMD/ terrorism and nonproliferation began frantically planning strategies. Just working with Iran to broadly monitor al Qaeda's WMD team would be invaluable, as would the eyes and ears the Iranians might provide in Afghanistan, Pakistan, and Iraq.

But then summer arrived and it seemed like no significant meetings were taking place. Rolf couldn't figure out why. It didn't make any sense. This was the goal: to get countries such as Iran to acquiesce to U.S. power and to come, hat in hand, to the table. It had been achieved, but then we didn't capitalize on it.

In a sit-down with the president at the end of that summer, Rolf asked why. Why didn't we engage Iran? He couldn't resist. Rolf was one of the country's key thinkers and actors on WMD. All this time, had he somehow misunderstood the actual intentions of U.S. foreign policy?

"It was one of the few times when I couldn't read his reaction," Rolf says, as he packs up his things to head to the State Department. "I asked about Iran, and waited. I was hoping to get a reaction. He just looked at me with this funny blank stare. I'd spent a lot of time with him, briefing him, and that had never happened."

Rolf downs the last of his coffee, now tepid, and begins talking, mostly to himself, running out the thread.

"By not taking action to engage the Iranians on at least the al Qaeda leadership inside Iran, we've created a situation where Iran now, in 2007, ends up being the most likely entity that would secretly be involved in strategic planning against the United States. And in a way, we allowed Iran to facilitate that, to end up in this position, by *not* talking to them. If there's an incident, like a WMD attack on the U.S., which is definitely a when, not an if, we'll track it to where it goes, and it's quite likely we would track it to Iran. And then Iran's toast, and we're toast. It's the worst scenario. It's the World War Three scenario."

Across town, Wendy Chamberlin is squinting up at the crown moldings of her new office. The place has a run-down, last-century feel, high-ceilinged but cramped, like a dowager's brownstone. She scribbles on the blank page of her new day minder: "renovations."

Chamberlin is a rare bird in this city—a lone flier, whom others are compelled to notice. In the past six years she's had about as good a tour as anyone for seeing how the world works. And the journey has changed her.

It's not that the players in Washington can't evolve. They can. But in this partisan era, participants in the great public debates often profit by staying put, safely on the side of one assembled team, even when their hearts may be ready to stray. Wendy—having traveled such a unique and wide-ranging path—is all but impossible for any team to claim. She's free to wander.

Which is what she's doing today, her third day as head of the Middle East Institute. It's an old-guard nongovernmental organization built by the "wise men" of the 1940s—many of them patricians—who guided America through World War II and gave the world the Marshall Plan. She has nothing on her schedule to speak of, but that seems to suit her, giving her time to get her bearings. After what she's been through, it might take a few days.

She left town in July of 2001 to become the U.S. ambassador to Pakistan, a job that seemed manageable, if challenging, after a twenty-eight-year career at the State Department that included tours in Laos, Malaysia, and Indonesia. As with every other American, nothing could have prepared her for 9/11. In her case—like a few dozen people in key jobs at that moment—being overwhelmed would have directly threatened national security.

She was there for many of the most important firsts: the first moments of startled clarity, the first phone calls from Washington to Islamabad, the first high-level meetings. On Thursday morning, September 13, she brought the list of eighteen key military demands to President Pervez Musharraf and sat stiffly in his office for forty minutes until he answered the question she'd carried from the president: "Are you with us in this fight?" When he said, "I am, without conditions," she got up and left. That night Powell called Musharraf briefly to run through the eighteen points that officially placed him at war with the radical elements who ran parts of Pakistan—including segments of its military and intelligence services—and with an organization they'd helped create, the Taliban.

Chamberlin visited America in the months following the attacks, for the first key sit-downs between Bush and Musharraf, which fused their personal relationship, leaving Bush feeling that the Pakistani leader was "a good man who could be trusted in every way." She was back at her post in January, for the first ritual beheading, the slaughter of The Wall Street Journal's talented Danny Pearl, which sent shock waves through the West and started a procession of similar slaughters.

And she was in her office in Islamabad on March 17 for the first suicide bombing in Pakistan since the mid-'90s. It killed an embassy staffer and her teenage daughter, along with six Pakistanis. She decided she had to order all nonessential personnel out of the country, which meant that her two daughters—twelve and fourteen—would have to go. As a divorced mother and their main caregiver, she would soon follow.

Not that her new job in Washington—running Asia and the Near East for the United States Agency for International Development, or USAID—took Wendy far from the fault line; that line seemed to follow her. The job controlled aid and development for a region that included Iraq, which she oversaw before the U.S. invasion and in the nine months following it. She can tell you stories galore about where the $3 billion she allocated ended up, things she did—built 3,500 schools and opened them on time—and things she might have done, which is a long list. When, in early 2004, Rumsfeld consolidated all such efforts under the "unity of command" at the Defense Department, and took the $3 billion, Chamberlin walked out the door and out of Washington. She was already fifty-four, but there was another job beckoning—something quite different. The UN needed a new Deputy High Commissioner for Refugees. So she packed up her girls, moved to Geneva, and ended up circuit-riding between the planet's worst pockets of human suffering.

There are many versions of misery but only two types of refugees: those who've been forced from their countries, a number that dropped from 12 million to 10 million between 2001 and 2006; and those who've been uprooted by strife within borders, called internally displaced persons, or IDPs—whose numbers rose from 5 million to 13 million in that same period. (Ibrahim Frotan and his family would have been categorized as IDPs in recent years in Afghanistan.) Put both groups together and you have roughly the population of Canada with nowhere to go and, in most cases, enough to eat and drink for only a few days. In a way, they might also be called non-state actors. The most resourceful of them end up illegally in some country, doing whatever's needed to survive in the shadows—prostitution, drug dealing, trading in stolen goods—while carrying a powerful susceptibility to radical movements, which afford a sense of coherence and community, a kind of invisible citizenship, mocking borders in a way that's especially appealing to a refugee or an illegal alien walking the streets of an unfamiliar land. The recruits of world jihad are often from this global community. Flipping though a list of young would-be martyrs among Muslim men in their twenties, you'll find many with a brother or sister, or a mom, who died in a refugee camp. Chamberlin toured many of them.

Oval Office to refugee camp—this happened to be Chamberlin's journey since 9/11: a wide loop, crisis to crisis, that she officially finished three days ago when she collapsed into this office in need of renovation.

"I don't think the American public understands how alienated the rest of the world is," she says, coming out from behind her desk and settling in a wing chair. "They don't really understand why the world is so damn alienated."

She talks about close friends she has in Texas, conservative guys, some with military backgrounds, and how "they think it's still the other guy's fault. And if we just can get rid of the bad people—kill them, take them out—everything will work out."

Chamberlin, meanwhile, after years in leading roles, implementing this administration's improvised strategies for dealing with a changing world, is moving in the opposite direction, backing away from America's long-standing faith in force and from its recent application of it. Not that she's squeamish, or a pacifist. Quite the contrary. She delivered the military orders to Musharraf and demanded his support on behalf of the president. She played a central role in trying to consolidate the United States' surprisingly swift military victory in Iraq, only to watch the country begin its slow deterioration into sectarian and tribal conflict—the kind of conflict she'd soon see everywhere as the refugee commissioner. No, Chamberlin earned her insights fair and square, walking the length and breadth of the world's strife. "Establishing control over territory using armies," she says, "making control our objective, is a losing game in this era."

She has plenty of company in this assessment, and even more in the effort to rewrite America's playbook for engaging the world. This imperative is all but defining public policy in the first few months of the year, from the *Journal*'s column on disarmament to countless symposia at the town's think tanks, both conservative and left-leaning, about how old strategies are not working and new ones need to be developed, the quicker the better.

What binds Chamberlin to a disparate collection that includes Rolf and Sam Nunn, Candace and Rob Richer—is that they all believe that the solutions will be found outside the formal channels of the American government. "What I've learned—and this is hard for me to say after twenty-eight years at the State Department—is that the answers will probably come from people-to-people contact, rather than government-to-government."

What that might be, she's not sure. "Driving the car, or in the shower," she says, "I find myself thinking more and more about the Peace Corps. The U.S. government hasn't really created a program that reached people, really reached them, since. We sent our best and brightest to villages around the world. They got the worms, they ate the weird food. It wasn't a program measured properly in the number of irrigation canals that were dug. No, it was an idea, an idea that reached the people of the world. It was simply that we, who have been given so much, care about you."

She talks about foreign aid and its worth, and how the United States commits less than 1 percent of its GDP to foreign aid, one of the lowest rates among developed nations. She shrugs. "I've worked in every area of foreign aid, handled every type, billions of dollars, and I can tell you that currently—at whatever dollar figure—it's not really working. When it's government to government, it mostly goes into the pockets of the wealthy and corrupt. We need a new idea. I ask myself what is it? I don't know. I just don't know."

This line of search brings her, quite naturally, to the Marshall Plan—the massive foreign aid expenditure, the greatest foreign policy act, arguably, of the twentieth century—and the reasons it worked so well. It wasn't undertaken to make our defeated enemies suddenly like us. No, the Germans' and Japanese' hatred of America was too deep-seated. We had killed hundreds of thousands of their civilians, women and children, in Hamburg and Dresden, in Hiroshima and Nagasaki and Tokyo. And it wasn't done so we could command lasting leverage over the defeated countries. The mid-century globalists, led by Marshall, moved forward because it was, as Wendy says, "the right thing to do, and when you do the right thing, you don't ask for anything in return. You do it because it's right, and because you can."

She lunges forward, almost out of the wing chair. "That—right there!—is actually a core American value," she says, and then goes into a five-minute speech about how that value means doing something because it's right, and we can, and how the recipients are free to do whatever such support empowers them to do, because we're supposed to believe in free will and self-determination—"those are core values, too." And, of course, the Japanese and the Germans eventually came our way and became our friends, when they were good and ready.

But would this work in the current context, with so many countries, or terror networks, or global ideological movements, that have declared themselves our enemies?

"I don't think we have any enemies," Wendy says.

And then she starts to laugh, this small blond woman, in her red dress, with a small gold broach and sensible shoes. After walking through refugee camps for a few years, she's ready to explode. Frankly, she can't believe she just said that—and she lets it hang in the dusty, sunlit air for a long moment.

"I think everybody, deep down, wants us to succeed, because they want the *real* values of America to succeed. You see, we don't trust our own values enough, that's the problem. People want the values we stand for. They want to be our friends."

Even al Qaeda? She says of course there will always be people strategically allied against us, but they know they can't challenge us, not really, not "if we're true to our oath."

She starts searching around for these values, these real, yet often inconvenient, values—incompatible with the current ideas of how power is used and augmented—and sketches outlines: "Marshall Plan values, Peace Corps values, and really believing in self-determination, free will, education, and opportunity—for their sake, not ours."

Right now, in symposia this very day at nearby institutes, there are heated discussions being had about how fundamentally untenable America's security situation has become, and will increasingly be, as technology—cursed and miraculous—guides destructive power on its downward cascade, all the way to the bottom, to the individual, the common man.

Our intelligence is thin, with knowledgeable people in the Islamist community shrugging and looking the other way; the cooperation from other countries, even allies, is often halfhearted. If the United States went down this path Wendy suggests—embracing a non-transactional, do-the-right-thing foreign policy—would people around the world start telling us what we needed to know before the moment of crisis, before one of our cities went up in flames? Isn't that the dark urgency beneath the term *hearts-and-minds struggle*?

Soon Wendy will be swept up in the quotidian swirl of plans and renovations, hirings, budget balancing, and the duties of leading this small institution with its sleepy stated purpose of bringing "greater understanding between the people of the Middle East and America." But, today, she can occupy the present, a space to exhale between where she's been and where she's going.

"If you think of hearts and minds as a policy, as a tactic, as a strategy to regain some of the moral authority we used to have, you've already lost," she says, feeling her way. "Because it's not a tactic, and it's not a strategy. The Peace Corps, the Marshall Plan, were things we tried that were real, real to people. Because, you see, they weren't about where we'd end up or what we'd gain, but about how we want to be in the world.

"That's the idea of a *value*, right? It's how you want to be, how you want to live. And that's what we need to transmit to the world."

She recognizes that she's thoroughly unsuited for the town's prevailing political conversation and that her ideas might not take seed anywhere in Washington's salty soil. "Sure," she says, "I'm a partisan, too. I'm in favor of *authenticity*."

I ask her, after nearly three decades in government, how these values—this longed-for authenticity—might translate into action.

There's no pause. "People to people. Great waves of us—just regular people, young people, building clinics or digging wells, laying electrical wires, carrying computers. Something, something big. America's good at building things. Let's do that, something we're good at.

"And ask nothing in return."

Chapter 2
What We Knew

Rob Richer is a can-do guy.

Can-do guys may end up running the world—and doing all sorts of things, no questions asked, in the name of America—unless they can be stopped first. And that's really an issue between the American people and their government.

But in making that decision, you have to understand how Richer ended up at the top of the can-do mountain, and further see why that mountain moved from inside the government to the land of the contractors.

A short version of how this happened would certainly note changes starting in the 1960s, when there was a concerted effort to bring government order, audit oversight, and quality control to a vast, varied souterrain of secret foreign operations being run by the United States.

CIA, springing in 1947 from the WWII Office of Strategic Services, had grown by then into a force of several thousand in the official task of clandestine operations. The agency was also paying thousands of foreign agents for their services and information. These assets and sources were sometimes officials of foreign governments or businessmen, and sometimes they were criminals and thugs. Of course, much of the money was wasted and plenty unaccounted for, but that was to be expected with this sort of enterprise. In the scheme of the vast U.S. budget, it was a penny on a hundred-dollar bill.

In any event, government did what government does. As order was imposed on this vast intelligence bazaar, some ugly practices were revealed, and this prompted a number of investigators in the early and mid-'70s. It seemed that duly elected officials, including but not limited to Richard Nixon, were periodically asking CIA to quietly do things that violated basic American values—attempted assassinations of foreign heads of state, for instance—and many U.S. leaders decided it was time to reconsider CIA's "by any means necessary" ethos.

By the late 1970s, the clandestine service was halved, only to dwindle further over the next two decades. Many of those remaining found themselves without much to do once the Soviet Union fell in 1991, and then more operatives and their bosses left during the poly-graph craze of the late '90s. It's hard to repair this sort of thing, to reverse course. The world's top intelligence services—the Israelis, the British, the Syrians—agree that it takes between ten and twenty years to build a top-drawer secret agent. But beneath all this attrition was a fundamental question: Can a democracy such as the United States—a nation that derives much of its per-suasive power from a public embrace of grand ideals—afford to have a robust intelligence arm doing things it might not want to admit to in public? The answer they came to thirty years ago was maybe not. All major op-erations needed to be briefed through Congress to make sure no laws were violated. Every action merited a pile of thick reports. The clandestine service that remained had largely become bureaucratized—ardent, but sleepy and slow-footed.

Then came 9/11, and U.S. leaders awoke into a frenzy of urgent improvisation, convinced they needed a large, skilled army of operatives to hunt down a new kind of enemy—terrorist networks—that had perpetrated the worst attack ever on the U.S. mainland. They en-listed people from every department—branding whole agencies, such as the FBI, as part of the intelligence business—and they hired a lot of new personnel, clean-scrubbed and ready to learn.

The government simply lunged forward. The new enemies were al Qaeda, first, and then a host of other radical groups, mostly Muslim, whom we termed terrorists; that was not to mention a growing band of rogue states, many of them in Asia and the Middle East, that may or may not have been supporting terror but were undermining U.S. interests in any number of ways: a grid of potential opponents charged white hot with the generalized fear of WMD. Flush with cash, CIA started to hire back the waves of agents who'd left the agency as contractors and found itself relying heavily on those few remaining members of the clandestine service with key contacts in the Arab world.

At the top of that list was Rob Richer, who, after several stints in the Middle East, had been stationed in Amman, Jordan, in the late '80s. Early on in the tour, Richer was asked, improbably, by Jordan's king Hussein to spend some time with the king's son Abdullah— a smart, free-spirited kid whose mother happened to be English, and who was therefore a long-shot for the throne. Richer, who has always managed to fuse entrepreneurial zeal with clandestine work, took to the task with gusto. By 1999, Richer—a former marine with a flair for quick thinking and quicker action—had helped guide the king's son into manhood. Richer was Abdullah's Karl Rove.

By that time, King Hussein was suffering from non-Hodgkin's lymphoma and several others scions remained favorites for succession. CIA was all over it. Jordan, after all, is among the most strategically important countries to the United States. With Israel on one side and Iraq on the other, it is, in essence, a large border town—population 6 million—that blends interests and cultures from several directions and has served as a rare honest broker in the troubled region. Hussein's brother, Hassan, had long been the crown prince and was expected to ascend to the throne. Another of Hussein's sons also had hopes.

But Abdullah—the dark horse—had what his competitors would soon complain was an unfair advantage: Richer, a man whose skills as a can-do operator even caused CIA colleagues to feel overmatched. It all came due when the terminally ill Hussein called Richer to his house in the Washington suburbs and told him that Abdullah would be the next king. Hussein's wife, the beautiful American-born and Princeton-educated Queen Noor, was kept outside the king's chambers as the news was delivered. So was Abdullah, who was soon summoned to the room. As Richer tells it—an account confirmed by a few others in the know—it was he, Richer, who delivered the news to Abdullah, with Hussein looking on, nodding his affirmation.

That night, Richer and the future king went out for a night of drinking at Nathan's, an old-timey bar-restaurant in Georgetown, celebrating a victory, as such, for both men. It would be Abdullah's last unescorted, unguarded outing—his last night as a civilian and not a future king. When the bartender announced last call, a rosy Abdullah turned to Richer and said, "I should probably get home and see my wife."

The king wasn't dead yet, but was fading fast, making for a very messy month or so of power struggle—a process that left Hussein in a state of apoplexy as he called then–CIA director George Tenet from his bed at the Mayo Clinic. "If you don't call off Richer," he shouted, "we're going to have a civil war." Tenet blew up, ranting around the seventh floor of CIA that he was going to fire Richer. And he might have if he'd had the time.

But Hussein died the next day, and at that moment, Rob Richer became unfireable, one of only a few truly indispensable employees of the U.S. government: the man who had helped create a king. Tenet, the former congressional staffer, resented it, while old CIA operatives just shook their heads. Richer had outmaneuvered the entire agency.

A few days after Hussein's death, Richer had an hour-long breakfast in the White House with Bill and

Hillary Clinton, briefing them on the character of the new king and what to expect from Jordan moving forward. He was soon leveraging his special status across the region—whether it was for CIA or for himself was a tough line to draw.

In the years since, Abdullah has been well advised, one might say, becoming a favorite of Bush while managing to retain currency and respect among Muslims in a strongly anti-American region of the world. Richer and Abdullah are godfathers to each other's children; they go on trips together and share every confidence. This special status makes Richer a true power player—one who left CIA, and the government, in the summer of 2005.

And that, of course, says a great deal. In an earlier time, the government would have done whatever was necessary to keep someone like Richer in the fold. Now, though, it has little credibility in making its traditional claim that, yes, you can leave and make more money, but you'll no longer be at the center of events. As the reliance on contractors has grown—now accounting for more than half the intelligence efforts in the so-called war on terror—Richer and his peers can do, for high pay, nearly everything they once did inside the government. In some cases, they're providing key functions the government is no longer able to perform. This combination—money and relevance—has proven to be unbeatable and surprisingly synergistic. The growing relevance of contractors to the basic functions of government gives them leverage to command ever increasing compensation, and—as a kind of proxy U.S. government—global credibility. Naturally, there is a federal system of competitive bids to control prices and enforce quality, but—as government accountants and oversight committees have discovered again and again—it is a system that repeatedly shows the advantage of "who you know" over "what you know."

Richer entered the contracting sector straight from CIA, becoming vice president of intelligence at Blackwater USA—the notorious private military company now since renamed Blackwater Worldwide—in the fall of 2005 and CEO of Total Intelligence Solutions, a spin-off company, in February 2007.

On March 20, Richer strolls through a mall atrium in Northern Virginia, all controlled affability, pinched, highly caffeinated, with a smile that makes him look a little like Regis Philbin. He sits on a bench in the airy, outdoorsy expanse of the mall's atrium, intelligence-dapper in his black turtleneck and blazer.

He's busy—CIA's station chiefs are in town—and he keeps in close touch with everyone who used to run the agency, including Rolf, whom he calls "the best big-ideas guy CIA's had in years." Richer's also busy with his new job, running operations all across the world, a few of which he describes in very general terms, mostly moving problematic individuals and large weaponry—such as helicopters and attack vehicles—across borders, finding people who don't want to be found, and helping a variety of officials in the U.S. government who regularly call on him. This morning Richer briefed a congressional intelligence committee about CIA interrogation practices—something he'd dealt with tangentially while at the agency—to give them the lay of the land. He revealed that the interrogations were videotaped, and commented simply, "It's ugly, not something you want seen by anyone."

But not all intelligence activities are meant to stay out of sight; one of the main goals of Rolf's "teams" is to confront the complacent country with the real threat of nuclear terrorism.

Richer's been talking to Rolf almost every week, and they saw each other a few days ago. Richer is excited. He's already got a team for the great uranium heist ready to launch.

He says he's thinking it should be off ledger—something he's doing on the side, and not inside the company he now heads. He says he's figuring all that out. But "we're ready to do it. I've got the operatives."

He explains that "he's using people with experience in Russia and that they'll run a few black operations," straight for-profit activities such as smuggling and selling guns, to build credibility, and then they'll go hunt down the uranium. "But what I'll need"—something he's told Rolf—"is a get-out-of-jail-free card."

"If I get two guys picked up in the 'Stans"—as in the former Central Asian Soviet states—"or wherever I get the stuff, they go to jail for life or they're executed. That's why I need the get-out-of-jail-free card." That, presumably, would mean the U.S. government intervenes, quietly, and says this was an elaborate test to discover the true nature of the current threat, or some such.

Rolf is the enabler—the guy, at this point, who can get Richer the needed immunity—an official letter showing that his operatives are working at the behest of the U.S. government. And Rolf can also write the check. Previously, Richer said he could pull a team together for as little as a million and a half. His "for-profit activities," however, seem to cover that, making his team self-supporting. But there are complex financial logistics with any potential seller, who would certainly want cash—up to maybe $10 million, Richer estimates—for enough uranium to construct a device. All these are things he and Rolf are talking through: operational matters. Do they, for instance, immediately capture the sellers and possibly drag them to the United States? Or do they set up an ongoing sting, paying for the uranium and then eventually tracking the seller to accomplices in a wider network, something that would almost certainly demand the involvement of other countries?

With each step, Richer offers a tour of the dark side. Some of his proposed actions, such as selling black-market weaponry, violate the laws of sovereign nations, just as espionage does, just as the "rendition" of uranium sellers to another country or the United States would. That's why he needs the Monopoly card.

But this slope is particularly slick. Would another country reasonably accept Washington's explanation about why a U.S.-supported team was running operations on its soil? Would it cede self-interest to the importance of the mission—a mission that might well create an international embarrassment? It might show, for instance, that Russia—to pick one obvious country—is as ready a market for nuclear materials as its critics charge, or that midlevel Russian bureaucrats are complicit in supporting a network. Or worse—that a sale is being secretly enabled by the current government. Same could go for Iran.

Certainly, the United States has run sting operations before. But both the needs and the stakes here are so high, as are the potential consequences of a startling disclosure. Constructing a team of people familiar with the dark side—and leaving them to do what's necessary to find out what's needed—is precisely what had created suspicion about the United States and raised questions about its commitment to its stated ideals of truth and justice.

It's Richer's view that people far and wide would understand that "the U.S. needs to do what it must" in this one area. What, for instance, if this off-ledger sting operation unearthed a network ready to sell uranium to al Qaeda? Beyond alerting the world to the true vulnerabilities of the "new nuclear age," it might actually uncover the most dangerous non-state players before they became history's actors.

But as Richer runs through some extralegal logistics, it's easy to understand how Rolf whipsaws between poles—between championing "a black ops" approach designed to boomerang hard truths back onto the United States, a "shock to the system," and sticking with a stay-the-course model of continuing to work, with all one's might, within an understandably broken system of public service. Maybe it can be made to work. Maybe the elephantine, self-defensive U.S. government—disintegrated and disintegrating—can learn to dance, to lead, and to stop paying hundreds of billions to contractors whose first concern is for the bottom line, even when they dive into life-threatening work. Maybe, at day's end, the great beast can self-correct.

But meanwhile, and until that day, Richer can provide access to Middle East leaders and a buffet of *can-do* services. The movement—across the past twenty years, and more with each passing year—to privatize as many functions of the government as possible, has found, in the war on terror, amber waves of grain, with several hundred billion dollars a year now being harvested by the contracting community.

Frankly, there's nothing Richer would like more than to startle the federal behemoth, proving he can do what the government itself can't—no questions asked.

Richer considers that an act of patriotism. It would also be very good for business.

It's the first week of spring—a moment of ephemera when Washington's cherry trees, a 1912 gift from the mayor of Tokyo, explode in pinkish-white blossoms. "If," as Emerson wrote, "we are reformers in the spring, but in autumn we stand by the old," it can be safely said that, across Washington, there's new growth—fitful, and maybe as short-lived, as the blossoms, but quite varied.

Wendy Chamberlin, it so happens, is polishing a speech on the "people to people" connection and it's promise, while Candace Gorman is preparing her next filing for relief for her client—both legal and medical—to the D.C. Circuit. Rolf and Rob are on the phone, thinking of ways to shock a slumbering America. And on the seventh floor of a glassy office building a block from the White House, a model of extra-governmental effort called the Nuclear Threat Initiative is busily carrying forward the idea that grim, apocalyptic facts are a blessing in any season.

The premise that a dangerous new nuclear age is dawning has become accepted around Washington. That has to feel like forward motion to Sam Nunn, NTI's cofounder. Within the graybeards, Nunn was something of a ringer. Over the last two decades, he'd established himself as one of the country's most respected and vocal champions of nuclear security—most recently as the head of this organization that tries to scare and dare governments into action.

NTI was founded in the spring of 2001 by Nunn and Ted Turner, when the two found they shared a nightmare vision of nukes in the hands of non-state actors. Nunn brought to the table long experience and respect in the policy world of national security; Turner, worth approximately $12 billion, had means and added an entrepreneur's flair to the mix, a passion for the big idea. Of course, they clashed. Turner, calling for nuclear disarmament, wanted quick, sweeping results. Nunn, the taciturn Georgia Democrat, knew from years in the policy establishment that there were limits to the scale and pace of institutional change. But their offsets worked, as did their timing; with NTI's founding coming just months before 9/11 showed the world terrorism's new face, the enterprise quickly took on a feel of prescience.

Nunn, however, had been at this for quite a while. He'd been fearful of "loose nukes" since the days of the Soviet Union's fall, when he was in a position to act. In 1992, he and his friend Senator Richard Lugar, the equally laconic Hoosier Republican, co-sponsored legislation to fund Pentagon efforts to secure the sprawling nuclear industry of the crumbling Soviet Union. This was the birth of the Cooperative Threat Reduction program, or "Nunn-Lugar"—which endures to this day and has neutralized thousands of nuclear warheads and hundreds of ICBM.

In those days, Nunn was impatient with calls for disarmament, thinking that the nightmare weapons were an ethical necessity—valuable deterrents in a delicately balance world. Disarmament, he felt, was a dangerous fantasy. But his career after the Senate changed him.

Since founding NTI, Nunn has seen with ever greater clarity how steep the odds of preventing a nuclear detonation are in a world awash with weapons and materials and murderous intent.

NTI's response is the application of steady pressure—pressure that, they hope, will yield steady progress, especially at a time when nearly forty developing countries, from the Persian Gulf to Africa to South America, are actively ramping up nuclear energy programs, building the tools and know-how to eventually, and maybe secretly, reprocess spent fuel into bomb-ready plutonium—the process the North Koreans successfully employed. Many of those among these three-dozen-plus countries are in the Middle East, including Bahrain, Egypt, Libya, Morocco, the United Arab Emirates, Saudi Arabia, and Syria. Even if all of them submit diligently to international inspections and are candid about where their desire for nuclear energy ends and that for nuclear weapons begins, the amount of uranium and plutonium floating around the world markets will rise precipitously. And these fissile materials will be residing in some of the world's most unstable countries, including those in the throes of a global Islamic revival. Even Persian Gulf countries with vast oil reserves say they have a right to tap the power of nuclear energy like everyone else, though no one doubts that part of what is driving this new nuclear age is the desire of many such countries to have a deterrent capability, and maybe even an offensive capacity. But they say otherwise—*no, it's only for energy*—and that winking denial is sufficient cover for Russia and China, and, yes, many companies in the United States, to supply plants and parts as fast as they can. This is the kind of profound global dysfunction that drives Rolf and Nunn to desperate actions.

And desperate innovations. It was Nunn who framed the guiding question for Rolf and others in his line of work: "The day after an attack, what would we wish we had done? Why aren't we doing it now?" NTI has done the best with what it has. The group's bankrolled operations that cleaned poorly secured HEU (highly enriched uranium) out of a civilian reactor in Serbia and down-blended thousands of pounds of weapons-grade uranium in Kazakhstan, on the Iranian border. The group's most recent and most ambitious initiative has leveraged funds provided by investment guru Warren Buffett to establish, with government cooperation, an international nuclear fuel bank. By ensuring an uninterrupted supply of nuclear fuel, such a bank would remove the incentive, or excuse, to develop domestic enrichment capabilities for countries that claimed to want nuclear energy.

All of these NTI programs are aimed at simply giving the forces of security the greatest possible statistical advantage. It's a probability game, trying to make sure the clear aim of terror networks to use WMD is a long shot—with longer odds every year—by consolidating nuclear weapons and materials in a few countries at few sites, and guarding them rigorously.

At least, that's the idea. Nunn is on the road today, where he spends much of his time, giving a fierce, dare-to-be-great speech, over and over, with all the energetic consistency of Harry Belafonte singing "Day-O." The man handling the day-to-day execution of the grand mission is Charlie Curtis, the head of NTI, who, on this lovely spring morning, is engaged in his most common activity: haranguing visiting legislators. The organization's staffers, meanwhile, putter about in the shadowy trenches, updating the widely visited NTI website—a gathering place for all matters of nuclear security—and cranking out polished, beautifully typeset reports designed to terrify the complacent. They are earnest and red-eyed, as if they've been staring into the sun.

This forty-strong team is a pound-for-pound fighting force as tough as any in Washington. And it needs to be, because not even here, in this realm of discernible, indisputable save-the-worldism, is the will to power—and its corporate corollary, the will to profit—held at bay.

On this score, the staffers tell a bracing short story about a prescient piece of legislation, the 1992 Schumer Amendment, that dealt with trade in HEU, used for nonmilitary purposes such as cancer treatment and imaging equipment. The act said that any company using American HEU had to get on the path to converting to LEU, or low-enriched uranium, which works just as well and can't be made into a bomb. The largest maker of medical isotopes—a $2-billion-a-year Canadian firm called MDS Nordion—paid lip service to the amendment for a decade while other companies made the conversion to LEU. Then Nordion, which had never made any significant efforts toward the shift, hired teams of lobbyists who started a letter-writing campaign to Congress and enlisted friendly scientists to overturn Schumer. Everyone lined up on the other side—experts in everything from medical imaging to nonproliferation—but Nordion and its allies carried the day. Legislation sponsored by Senator Richard Burr, a North Carolina Republican, was tucked into the pro-business Energy Policy Act of 2003, which exempted Canada and four other countries from Schumer's provisions.

Senator Chuck Schumer, the New York Democrat, commented that it was "amazing we would do this in a post-9/11 world." It's a common refrain of politicians that everything changed that fateful day, but this story—told by the still-wild-eyed NTI staffers four years later—underscores all that hasn't. The entire matter of Nordion and its uranium fits with a grim understanding of the way power—whether economic or political—is exercised with few checks or accountability. It gets what it wants.

Charlie Curtis wanders into the conference room and slumps into his chair. NTI's president and chief operating officer fights back resignation. "All of what we do here is just a question of time," he confides. "We are trying to buy time."

He means that NTI is doing what it can to hold things at bay, but that its efforts do not represent a permanent solution, and without a full reinforcing architecture, he and NTI are just postponing the inevitable. He's not optimistic that the government is doing much to build this reinforcing framework.

"We've gone from treating these issues—securing weapons materials at their source to the highest possible standard to deny theft and diversion—from important to very important, but it's still not urgent. Despite what the president says, the vice president says." There's weariness in Curtis's voice. "The political leadership of this country, those in office and those contesting for office, all believe they have this message—they all believe that nuclear terrorism is their number one priority. They all understand the transforming effects of active nuclear terrorism."

But little is being done. "We're puzzled why our government and other governments aren't doing more with a higher degree of urgency, and it's confounding," Curtis says. The issue has been acknowledged and parsed at the highest levels, but our leaders continue to fail to answer Sam Nunn's simple two-part question: "The day after an attack, what would we wish we had done? Why aren't we doing it now?"

For Curtis, it's a matter of seeing "this as an existential threat to the *nature* of this country *as well* as its security. It's not a matter of just losing blood and treasure."

Curtis is not optimistic about a future that promises larger and larger power to smaller and smaller groups. On the sore subject of the Canadian company, he shakes his head, saying he hopes that maybe we will come to define shared interest more broadly and that the public can become a more "insistent conscience in the political process."

The Nordion fiasco may seem a modest defeat in the grand battle to protect nuclear materials, but it is a precise emblem of a government, in a time of partisan warfare, not being able to act coherently—to be integrated, as Plato might suggest—around even the most basic shared principles.

The dilemmas, though, are dense. The "insistent conscience" that Curtis yearns for is nourished by frank public dialogue, sustained by at least a few clear, discernible facts. Nordion, in its well-funded efforts, worked ardently to cloud such clarity, which then gave legislators just enough cover—much like the Russians building nuclear power plants across the Middle East and Asia—to wriggle free from the consequences of their actions. The goal, power's goal, is subtly to corrupt the heart of accountability, with an eye toward allowing actors of all stripes to do what they please for whatever reasons they decide.

Now add another feature of what might be called liberating complications: the reflex secrecy that cloaks almost everything in the fight against terrorism. If the public knows, the thinking goes, so will the enemies of the public—the terrorists—and who would want to hand them that advantage? Yet, the flip side is little noted: the horrifying fact that the participants themselves in these public debates—the legislators, lobbyists, companies, and government officials—may not know fully what they have done. In this case, Nordion may feel it knew everything, including a few things that it would rather others not know, such as the troubling fact that its source of HEU, the Chalk River nuclear laboratories near Ottawa—a nonmilitary facility—housed enough nuclear material to make a bomb.

What Nordion did not know—nor did Curtis or Nunn, or Burr, for that matter—is that the intelligence community for the past five years has been deeply concerned about a rather large and growing radical Islamic community inside the large, varied Indian reservations that line the Canadian side of the long and porous U.S.-Canada border. The subject has prompted screaming matches between American and Canadian intelligence officials, with the United States accusing Canada of being timid in not pressing for more electronic surveillance and infiltration of these communities, and the Canadians responding that they are operating within their laws and, unlike the United States, not only when it's convenient to do so.

Who profits from this fact being secret? The Islamic communities on the U.S. border certainly know that they've been noticed. The government, of course, gets to control what goes on behind the scenes and what is made available for the public to assess its performance. Legislators act on inadequate information, the companies profit, and if there's a catastrophic event, everyone loses.

It is of course the same connect-the-dots gremlin that was visible in the reviews and recriminations after 9/11: the inability of almost anyone—citizen, legislator, company chief, even intelligence officer—to see the whole picture.

And, in fact, the whole picture gets wider still.

CIA knows these Indian reservations well, has known of them for a long time. It's where the Soviets used to drop their spies—an ideal place, then and now, to walk into America.

The terrorists *in the Miami warehouse work through the final stages of the construction of a dirty bomb— an explosive device wrapped in radiological material that'll scatter at detonation.*

It's a sweltering day in mid-June, and two SWAT teams, one from the FBI, one from the Miami Police Department, edge toward the warehouse—men sweating in fifty pounds of gear, guns poised, moving in a tight formation.

Suddenly, in a swift coordinated movement, both teams race toward the warehouse's entrance and bust their way inside, guns blazing. The terrorists return fire, some are hit; in a moment, all are captured.

But that just brings the teams to their most difficult step: safely disarming the bomb and removing the radioactive materials.

From on high, one Russian intelligence officer says to another, "Where do they get those helmets?"

"The football team, okay?" his partner says. "The Dolphins must have left them behind in the locker room."

The two men share almost imperceptible smiles over this, as they slouch in their blue seats at the Orange Bowl, along with three hundred other hard-to-impress-but-happy-to-be-here law enforcement officials from twenty-eight countries.

This is the signature day, the fun third day, of the weeklong "Global Initiative to Combat Nuclear Terrorism Law Enforcement Conference," sponsored by the FBI for the world's police and intelligence elite, or the officers working just below the elite, and maybe a few, or more than a few, in the dreary middle rungs, and plenty who could just use a few days of sun in Miami.

And sun they get. And free Pepsi today, or bottled water, care of the FBI. From the announcer's booth, a color commentator who's narrating the action on the field—"a mock drill to show how we'd respond to a threat involving a weapon of mass destruction"—tells the spectators, bused in this afternoon, June 13, from conference headquarters at the Miami Intercontinental Hotel, that there'll be a short break before men in hazmat suits simulate the disarming of the bomb.

The assembled mill about and take in the scene: the famous stadium with its end zone palm trees, once home to Miami Dolphin Super Bowl feats and college football spectaculars, now hosting a different kind of staged combat, with fewer frills and surprises. The "warehouse" that the SWAT teams run through is just low storm fencing, set up at midfield to mimic the floor plan of a building. The nightmare device is a small cooler on a picnic table.

If, as the 9/11 Commission concluded, America's vulnerability to a terrorist attack was due to "a failure of imagination," the Miami conference is one type of response: a few days with speakers and exhibits and today's mock drill that are designed to nourish imagination and morale, to help the five hundred or so attendees see that they're actually soldiers, warriors, in a great global battle—a battle against something that's never happened before.

It's a tough task and a tough audience—men, mostly, with concrete apprehend-and-arrest sensibilities. And they don't scare easily.

Still, the speakers did their best, starting with FBI director Bob Mueller, who opened the conference two days ago with the warning that "there is enough highly enriched uranium in global stockpiles to construct thousands of nuclear weapons. And it is safe to assume that there are many individuals who would not think twice about using such weapons. The economics of supply and demand dictate that someone, somewhere, will provide nuclear material to the highest bidder, and that material will end up in the hands of terrorists."

Fine, the five hundred nodded. So what do we do? Whom do we go get?

No one, specifically, of course. But Mueller, like virtually every speaker over the next two days, ran forcefully through the general array of worthwhile actions—intelligence sharing, training programs, nuclear accounting and housekeeping—in speeches that were piped in to a companion conference the Russians were heading up in Kazakhstan for officials from that region who couldn't make the trip to Florida.

The "Global Initiative," after all, is what Putin and Bush signed the previous July at the G8 Summit in Russia. This, so far, is the outcome: conferences. And this conference is the biggest yet.

Early this morning, before everyone boarded the buses for the Orange Bowl, the man behind all of Miami's sound and fury sat at a table in the Intercontinental's lobby. His name's Vahid Majidi, the FBI's assistant director for the Weapons of Mass Destruction Directorate. He's a chemist, most recently chief chemist at Los Alamos National Laboratory, and the FBI's representative to the IND Steering Group that Rolf chairs.

He's also the highest-ranking Muslim at the bureau—and among the highest ranking in the U.S. government—which has an astonishing paucity of followers of Islam in its upper reaches. Majidi came to America as a high school student in 1979, when his parents fled during the fall of the Shah. He went to colleges in America, taught chemistry at a few, and now heads the FBI's efforts in the most contentious area of America's relations with the Muslim world. His daughter, a precocious ten-year-old, recently asked him if he thought this was "ironic."

"So I say to her, how do you mean?" he said, recounting the conversation.

"Well, you being born in Iran and now running the WMD program."

"I said, 'To be honest with you, I never thought about it that way. For the past ten years, I was helping to make nuclear weapons. Now I have to make sure we stop them.'"

Not much on irony, that Majidi. He understands, though, how convenient it would be for him to act as an emissary from America to the Muslim world. But it rubs him wrong: he's a scientist and wants to be judged as such, in the classic émigré model. The compromise is that he's recently started to speak to groups of scientists in Muslim countries, talking the lingua franca of science. Bin Laden has been openly recruiting physicists and chemists, even some biologists, too, and there's evidence that anti-American sentiment and nationalist feeling inside Islamic countries—or simply among many Muslims who believe they are being attacked by the West—have nudged Muslim scientists and Muslim radicals toward each other. Indeed, some of those whom Majidi meets of late hold extreme beliefs. But, he said, "those ideologies drive the ultimate application, not the initial discovery." He engages them because they're colleagues, and all scientists love talking science.

"If there is a fundamentalist view," he acknowledged, "it doesn't really matter what I would say. It wouldn't change instantaneously. You have to have a dialogue over a very long period of time. The only way you can chip away at established lines of thinking, established fundamental views, is by a long-term dialogue. And chip at those things a little at a time."

In the meantime, the "markets" remain active.

He mentioned the Russians and their ongoing obstinacy about accounting for materials and disclosing smuggling incidents inside their borders. "The Russians have lots of priorities," Majidi said. "This may not be at the top of their list."

More broadly, he explained, "One of the big advantages we have is that we have a market that is ambiguous. . . . Say you have material. Who are you gonna try to sell it to? And how are you gonna find out if the person who wants to buy it is for real? We have a supply that's ambiguous. And over the past three decades there have been a huge number of scams that have dominated the market. So you have a transition that's ambiguous. So what we have going for us is an overall uncertain marketplace that no one . . . *no one* is a strong word— that many people have lost in. So it's an incredibly risky proposition. So that ambiguity has helped over a long, long time. It has given us some opportunity to sit back and try to find out what is the best way to get engaged."

Markets evolve, however, by their very nature, to squeeze out ambiguities. He acknowledged that this was happening, as Mueller had in his speech. "That's what this conference is all about," Majidi said. "We're trying to tickle one element of the market, to get to the countries that the material may start from" and "give them the tools and technologies, the investigative techniques, the intelligence" to reduce the amount of real material that is flowing into these "ambiguous markets."

And why would they want to cooperate, especially if they're probably not going to be a target of one of these catastrophic events?

His pitch—the American pitch—is fairly weak. "The idea is that we'll provide these countries the intelligence they need," Majidi said, summing it up, "and we're asking them for a reciprocating activity."

Later that night, another FBI man, one who'd spent

years in Russia, the Balkans, the 'Stans, assessed Majidi's pitch—FBI's pitch, really, for the past year—and how countries are responding to it. "People don't realize in America how little underlying credibility the United States now has in the world, especially on this matter of WMD, which, of course, has been driving everything," the guy, a conservative Republican, said over drinks. "We went to war—the most important thing a country does—based on WMD, and we were wrong. That's means either we're amazingly incompetent or we lied. Take your pick. Now, I think we lied—most people do—because no one could be that incompetent. But until we come clean—and here we are years later and we don't even care enough as a country to figure out what really happened—we're sunk. Every time there's some media report about Iraq, which is every day, people all over the world say, 'right'—lied or incompetent. That leaves us reduced. And it's sad. We're now seen as just like everyone else, a country on the hustle. Right here, in Miami, you can see it—how hard it is to lead these countries without real moral authority. It's like herding cats."

Back at the Orange Bowl, as a robot disarms the dirty bomb, rows of big cats—Ukranians, Russians, Georgians, Kazakhs, and Turks—sit yawning. There are surprisingly few representatives from any Middle

Eastern countries. But, of course, those are the homes of the buyers.

These large men represent the supplier countries, many of which are a mess of public corruption, crime, poverty, and cool attitudes (or worse) toward the privileged United States.

An FBI agent standing nearby who knows a bit about Lovebug—the Kurdish source who was one of CIA's few assets inside al Qaeda, and about the leak that got him killed coming from inside Turkish intelligence—warns against fraternizing with them. He points to a row of Turks, four men wearing black leather jackets, even in this heat.

"They're killers," he says. "A different breed."

But, then, they all are—*different breeds.*

Looking across the array, row after row, it's painfully obvious that all the best intentions of Majidi and his boss Bob Mueller will be for naught until America manages to recover some of the status it's lost: the nation of immigrants—not one breed, but many—that always did its best to uphold the highest ideals. We don't always make it. But we always try. And when we've messed up, we've done our best to get to the truth: to admit we were wrong, whatever the consequences, and try to make things right. That combination of effort and character is what underlies Lincoln's prayerful designation of America as "the last, best hope on earth."

Because you can, actually, herd cats. They can't be forced, of course. But if they sense something they want, if they're enticed by something good, they'll follow, even in herds. America has done it many times when, as Wendy Chamberlin said, it has been true to its oath.

And it was clear, suddenly, that there actually was one man, at this grand Floridian fest of futility, who was oddly pertinent to everything that had unfolded.

The fat man by the ficus. He was standing, inconspicuously, in the lobby on the first day of the conference. His name is Alan Foley, and he's now associate director for national security at the Argonne National Laboratory, a Chicago-area backwater of DOE.

Foley was once at the center of the action, head of WMD analysis for CIA, and generally considered the missing man in one of the great controversies of this period: the inclusion of the sixteen words about Saddam Hussein seeking to buy Niger's uranium in Bush's 2003 State of the Union speech.

It's one of a nest of linked controversies surrounding prewar intelligence that, together, have greatly damaged America's standing as an honest broker on the world stage.

Foley all but vanished since he left CIA in late 2003.

But his improbable presence here, a man from the past discussing the future, hints at a tragic irony underfoot and fast emerging: that America's misbegotten WMD case for war in Iraq has undermined its moral standing on a mission it now so desperately needs to lead—this one against a real threat: nuclear weapons in the hands of terrorists.

Foley was never interviewed on the record at any length in the aftermath of the sixteen-words scandal, though many have tried. Not that the public statements of those sitting in key chairs during the debate on prewar intelligence have been particularly illuminating. Despite the obvious value of candor in restoring trust in the United States, and after inquiries by a senate committee and an independent commission, the key players, from the president on down, have said surprisingly little on the subject—surprising, considering how vocal many of them were in the run-up to the Iraq war.

The case for war, at day's end, broke down into two parts: Saddam's nuclear weapons activities and his program to develop biological weapons. Of course, nothing, ultimately, was found in either category. The outstanding issue is whether this was the result of willful misrepresentation by the U.S. government (a lie) or profound ineptitude. The position of U.S. leaders is that they were told by intelligence officials that their assertions in both areas were sound and that there was no evidence to the contrary. In short, it was an intelligence failure by those who served the president and vice president.

Foley's role becomes integral in the first part—the nuclear case—which was founded largely on an intelligence report about Hussein trying to buy yellowcake uranium in Niger. For all the attention this has received, the question endures: How did something CIA relentlessly opposed end up in the president's address? Here is a brief synopsis of what is not in dispute: An Italian document claiming that Hussein had tried to buy the yellowcake was passed to both British and American intelligence in 2002. The British intelligence believed the information was basically sound. CIA did not. One reason was that Saddam already had five hundred tons of the stuff, which can be made into bomb-ready material only with a long, expensive, hard-to-conceal manufacturing process. Why would he want more?

But the government was slowly breaking apart by the summer of 2002. The Defense Department and its intelligence unit, called the DIA, both under the sway of various top officials—such as Paul Wolfowitz, Stephen Cambone, and Douglas Feith—were anxious to give the president the evidence he needed to declare war on Iraq. In the classified National Intelligence Estimate, or NIE, issued in the fall of 2002, the Niger report was relied on to support the conclusion that Saddam was "vigorously trying to procure" nuclear material. NIEs are supposed to be consensus reports of all the nation's intelligence agencies, which distill best judgments through an interagency process. That didn't seem to happen here. Foley expressed surprise that the Niger report and resulting conclusion had found their way into the estimate, and he called in his analysts, who said they were never consulted on the NIE. Though Foley didn't know it, CIA director George Tenet had successfully had a phrase almost identical to the sixteen words expunged from a speech the president gave in Cincinnati in October, after a tense conversation with National Security Adviser Stephen Hadley.

The issue reared up again a few days before Bush's 2003 State of the Union Address. Bob Joseph, a White House specialist on WMD and proliferation issues inside the National Security Council, called up Foley. Joseph said he was working on the State of the Union.

Foley, in the hotel lobby, recounts their conversation.

"The original text that Bob gave me said, 'We now know that Iraq is seeking yellowcake from Niger.' And I said, 'Bob, don't say that.' . . . His point was, 'It's in your estimate. It's in the NIE.' And then I realized, you know, Bob's sitting there with the NIE. He was saying that we, meaning CIA, had this view. The NIE view was DIA's view . . . I found that out after the fact. I had my head analyst in. I said, 'Where'd this come from?' He said, 'Well, DIA.'

"So Bob says, 'You know, it's in the estimate.' And I said, 'Yeah, but that's also classified, Bob. This is a classified judgment. I don't want the president going out there and saying in the royal *we*, "We now know." 'Cause you're basically pointing your finger out here at Langley, saying, "U.S. intelligence has concluded." Don't say that.' "

Foley recalls that they then discussed the topic of classification. "And then the issue of the British paper came up. My recollection is Bob brought it up. He said, 'It's in the British paper. That's unclassified, right?' His recollection is I brought it up . . . the 'dodgy dossier' . . . And so Bob said, 'If we said, "According to the British government, Iraq is seeking uranium in Africa," would that be factually correct?' And I said, 'Yes, that would be factually correct. It's self-evident. It's out there. It's done.' And we just sort of left it there."

That turned out to be the only consultation between the White House and CIA on the matter—one of the most important and contentious things a president, any president, has said in the State of the Union Address. In the months following the address, it became clear that the Italian documents were forgeries. There was nothing to the Niger claims.

"I didn't assume that whatever Bob and I agreed upon was going right into the president's desk in the Oval Office," Foley adds. "He has his own chain of command there. Hadley's in the loop. Hadley supposedly had to chat with Tenet about Cincinnati. How come Hadley didn't intervene? So it's an interesting question: How come these guys who were working on the Cincinnati speech didn't know that Tenet objected to the language in the first place? And even if they did know, why would they accept Bob Joseph saying, 'Oh, it's okay. I talked to this guy Foley out at CIA'? Does that make any sense to you?"

Over the next half hour, Foley begins digging to the very core of what occurred.

First he says definitively that there was pressure on all the analysts, something that no investigatory commission has been able to establish. "I got phone calls— I'm sure Bob did—not just from the administration, from Congress, saying, 'We don't want this *possibly, probably* shit. We want to know, what's your best judgment? What's your bottom line?' And the NIE fell into that trap. The famous sentence: 'Iraq has biological and chemical weapons.' We didn't know that. *We didn't know that.* We believed it. I'm not saying it was a lie. There's a big difference between 'we believe that' and 'Iraq has.' Why are we talking this way? My answer to that is we were under a lot of pressure to dispense with the qualifiers and say, is it black or is it white?"

More broadly, Foley says, dressing up this thin speculation as evidence was, at its heart, dubious. "That was one of the things that really got confused. This isn't a court of law. This wasn't evidence by any standard a good defense lawyer couldn't get thrown out. This is intelligence. . . . A big chunk of it is hearsay." He adds, "It is, in my opinion, true that the administration, for whatever reason, was determined to have a showdown with Iraq that predated this whole WMD stuff."

But it is clear that interpersonal issues may have, at day's end, made a great deal of difference in the matter. Foley says he and Tenet rarely saw each other. Tenet had his group of top deputies and aides on CIA's seventh floor, and Foley was never any part of that. "I wasn't the first one in his Rolodex."

He and Joseph, on the other hand, were quite friendly. They spoke two or three times a week, Foley says, and met with regularity. "Bob Joseph and I worked together *all the time.* I was on his committee. I was George's representative to the interagency group. And so for Bob to call me up and say, 'Hey, I want to say this in the speech. What do you think?' was perfectly normal."

Within this context, now finally clarified by Foley, Bob Joseph's decision to make an unremarkable call to a friend who rarely saw Tenet or his seventh-floor disciples—and not tell this friend that he's the last check in one of modern history's most contentious phrases before it ends up on the president's lips—was a perfect way to get something CIA doubted, and its director had already opposed, into the big speech.

What's clear is that there was already ample evidence inside the White House to know that the Niger case was extremely problematic.

Foley still doesn't blame his friend Joseph, one of Washington's savviest political players, or impute any ill motives to his call being made. But Joseph, Foley recalls, made it clear that "he was the policy maker and I was a fact-checker." What Joseph understood was context: where the Niger claim fit in the context of discernible truth, and where Foley fit in the context of the CIA.

That allowed Joseph, as the White House's representative, to know that Alan Foley was the only man to call.

The second part of the case against Hussein that led to war involved the biological weapons facilities.

The central source for this intelligence was Rafid Ahmed, an Iraqi defector to Germany better known by his codename Curveball. We know a great deal about him now. It is common knowledge that Curveball turned out to be a skilled fabricator and that his tales about germ factories and mobile labs—the core of Colin Powell's UN speech that made America's case for war—were false.

Much has been made of how the U.S. government, and CIA specifically, wanted to question Curveball directly but was denied access by the German government. Had they been given access, they might have discovered that he was a fabricator.

That's actually a cover story.

The truth is that someone inside of CIA—a man the Germans trusted completely—recommended to the Germans that they never provide the United States with access to Curveball.

His name is Joe Wippl. Another missing man in a picked-over drama. The key man.

Starting in 2001, Wippl was CIA's chief of station for Germany. It's a big job. Germany is the most powerful country in Europe, and Wippl, one of the best German speakers in the agency, handled everything German.

But by early 2002, it was clear inside CIA that there was a problem with Wippl. He was popping up in various European cities without following basic agency regulations that required him to report his movements to his home office and the local CIA offices in places he'd turn up. Large CIA stations, such as Vienna, would call Langley and wonder "what the hell's Joe Wippl doing here?"

He was also showing up in electronic surveillance the United States was conducting on various German officials, including some top intelligence officers. They were talking about Wippl and the things he was telling them. Several high-ranking CIA managers became concerned, including Stephen Kappes, the number two official in the Directorate of Operation, the agency's clandestine service; Richer, then CIA's head of Human Resources; and Tyler Drumheller, head of the European Division. They started an informal inquiry.

According to several of those involved, they found that Wippl's activities presented a particularly problematic mix. He was a married man—his wife lived in Northern Virginia—moving secretly among what seemed like several women, any number of whom could be German agents; he was spending inordinate, and unreported, time with Ernst Uhrlau, head of the German intelligence service; and he was telling his German counterparts things that were unauthorized, highly critical of the current CIA leadership, and sometimes at odds with U.S. policies, including—importantly—that the Germans keep Curveball to themselves. What also became clear to CIA electronic surveillance at this point, in mid-2002, was that the Germans felt Curveball was possibly unreliable. Wippl, they found, was fearful of placing an unreliable German source in the midst of the Bush administration's march to war in Iraq. He knew it would end badly and might cause a rupture in U.S.-German relations, which he felt was the most important priority, so he secretly recommended the Germans never give the United States access to Curveball.

"A lot of us wanted to fire him," said one top official involved in arguments about how to handle the Wippl mess, "but the very top guys didn't have the guts." Instead, Wippl was called to Langley in August 2002 and told he'd be leaving his post in the coming year. He was not told everything that had been discovered, because officials feared they would spoil the source inside German intelligence who had been providing them with information, counterintelligence, on Wippl.

At this point, the White House was ramping up its program to sell the war to the American public, and its case would rely heavily on Curveball. In the months leading up to the war, CIA officials to tried get access to the Iraqi, but they were blocked by the Germans, who were following Wippl's advice. But there was another player in the mix: DIA, which was closely aligned with the Vice President's Office. Defense Intelligence had managed to claim the Curveball matter as their own, and this meant that CIA's questions about Curveball and matters of access all had to pass through DIA.

The United States never got access to Curveball. The testimony he gave the Germans was filtered secondhand through DIA and became the core of Powell's testimony on February 5, 2003. Six weeks later, the United States invaded Iraq.

By late spring, the mobile trailers Curveball had described as at the heart of Saddam's biological weapons operation were found to produce hydrogen for weather balloons. They literally produced hot air.

U.S. troops, at that point, were in Baghdad. Joe Wippl, meanwhile, was essentially AWOL, moving through Europe and among various girlfriends, occasionally showing up at conferences representing CIA. He was still in close contact with German intelligence. At one point, he called a senior diplomat in Germany on what both men knew was a tapped phone—with the Germans listening in on the other end—and seemed to pass some messages along for those eavesdropping. The official, fumbling, told Wippl "this is not a secure line

278 • RON SUSKIND

and you know it," and hung up. By summer, Wippl was gone.

The perfidies of prewar intelligence were just then filtering out. Joe Wilson, a former State Department official, led the charge, publicly disputing the Niger claims only to have his wife, Valerie Plame—an undercover CIA agent handling WMD who once worked for Rolf—outed by administration officials. A yearlong bout of recriminations and leaks followed as it became increasingly clear that there were no weapons of mass destruction in Iraq. The White House blamed CIA, saying the agency was at fault for the specious intelligence behind the case for war. Much of CIA's management was gone, or on its way out, by the late summer of 2004. Tenet led the way and was replaced by a close ally of the White House, a Republican congressman and former CIA agent named Porter Goss.

And then a strange thing happened. The Vice President's Office recommended that Joe Wippl be hired for the plum job of congressional liaison from CIA. He was. Some old hands still at the agency speculate that Wippl—who was in regular touch with DIA officials at the time he was recommending to the Germans that they deny CIA access to Curveball—was secretly acting, in 2002, at the behest of the Vice President's Office.

Wippl, who was also friendly with various officials around Goss, including Dusty Foggo—who was later indicted on bribery charges—has admitted to friends he was terrible at the congressional job, but delighted to get it.

All the former directors of CIA, or DCIs, met at the agency's Langley headquarters in August 2005. It was a long line, including George H. W. Bush, Stansfield Turner, Bob Gates, Jim Woolsey, and Porter Goss. Goss would be the last of them. The CIA, injured—some would say mortally—by the struggle over prewar intelligence, would cease to exist as it had since its formation in 1947. It would now be subordinate to the Office of the Director of National Intelligence.

Wippl was fired as head of CIA's Congressional Liaison Office in 2006. But, somehow, he remained on the CIA payroll—and got another plum job, this time running a one-man intelligence "institute" at Boston University. For a man twice fired, it was astonishing good fortune. Somebody up there must like him.

There's a third short précis to offer in addition to the tales of Alan Foley and Joe Wippl: the story of Sabri.

Naji Sabri was Saddam's last foreign minister—exactly the sort of asset the White House and the intelligence community had been looking for in those months prior to the Iraq war. Having spent more than a decade living and working abroad in Europe, traveling on behalf of the Iraqi government, Sabri spoke near-perfect English and had developed an affinity for Western society and culture. Along the way, he had also established a relationship with French intelligence as a paid spy.

Though there were surface tensions between the United States and France in those prewar years, the countries' intelligence agencies maintained a good working relationship. CIA's Paris station chief, Bill Murray, was one of the more experienced field bosses in the clandestine service, having run five stations across a thirty-year career. The French liked and trusted Murray—a rough-hewn, brutally frank character from South Boston—and put him onto Sabri in the summer of 2002. Back in Washington, Bush, Cheney, and Rice were briefed on the development, and agreed that Sabri seemed very promising indeed. Langley concurred, and coughed up an initial payment for the high-ranking Iraqi: $200,000.

There was still the matter of arranging a meeting. The tense atmosphere of the time and Saddam's dictatorial paranoia meant that Sabri couldn't leave Iraq without good reason and an airtight cover story. Finally, in September, he had an excuse.

As foreign minister, Sabri had been engaged in ne-

gotiating the return of UN weapons inspectors throughout the summer of 2002; in mid-September he was due
to travel to New York to finalize these negotiations and
address the UN General Assembly on Saddam's behalf.
Direct contact was too dangerous. Arrangements were
made for Sabri to meet with an intermediary—a Lebanese journalist trusted by both sides—while in New
York. The intermediary would pose questions on behalf of CIA and then follow up with Murray.

The plan went off smoothly: Sabri passed along what
he knew, Murray debriefed the journalist in a New
York hotel, and for his address to the General Assembly, Sabri even wore a specific type of suit requested by
Murray as a good-faith signal.

The upshot of Sabri's account was that Saddam neither possessed WMD nor was trying very hard to procure or develop them. If Saddam was eager for a nuclear
weapon, he was as far as ever from having one and was
making no progress on that front; any vestige of a bio-
weapons program was negligible; and if any chemical
weapons remained in Iraq, they were no longer in the
hands of either Saddam Hussein or his military.

Murray flew down to Washington to deliver the news
and briefed John McLaughlin, CIA's deputy director.
McLaughlin was enthusiastic about the intelligence
but pointed out that it was contradicted by information

from Curveball, the best source on Iraqi WMD to that point. Sabri's account was relayed to Tenet, who delivered it personally to Bush the following day.

But the administration quickly lost interest in Sabri when it heard what he had to say. Bush dismissed the intelligence as disinformation, and the White House said it would be interested in Sabri only if he chose to defect.

Inside CIA, however, a number of officials were not ready to give up on such a well-placed asset and went back over the file to see if they could build a better case. The French helped by monitoring Sabri's calls, which were sent to CIA and which backed up what the Iraqi had revealed. Everything seemed to fit. Even contextual details added to Sabri's plausibility: for example, the fact that two of his brothers had been killed by Saddam helped explain both his willingness to cooperate with the United States and his caution in doing so.

While this was going on, CIA was putting together a report on the Sabri intelligence. Murray had been in a rush to get back to Paris after briefing McLaughlin, and he didn't stick around to write up the report himself. Instead, he passed the job off to the CIA station in New York, where he was connecting on his return trip to France. What eventually emerged from the New York station proved to be a serious distortion of Murray's initial filing. Most strikingly, a new introductory paragraph had been added that claimed not only that Saddam possessed biological and chemical weapons, but that he was "aggressively and covertly developing" nuclear weapons. These assertions were not just unrelated to Sabri's disclosures and Murray's reporting; they were in direct contradiction to both.

This erroneous report—almost certainly altered under pressure from Washington—was guarded so closely that it was never shown to the teams, at CIA and elsewhere, hurriedly assembling the October 2002 NIE on Iraq's WMD.

After further manipulation, the report was, however, deemed suitable for our foremost allies. An unsourced version was passed to Sir Richard Dearlove, Britain's top intelligence official at the time, and he notified Blair. The version of the story Blair heard was a series of square facts divorced from evidence, the first of which concerned Saddam's aggressive pursuit of nuclear weapons. Blair took this at face value.

It was quite an amazing game of Telephone to turn the disclosure by Iraq's foreign minister that the country had no WMD or serious WMD programs into a report, delivered to the British prime minister, that Iraq was vigorously pursuing nuclear weapons. Though the British and American intelligence services sometimes disagree, poor communication is usually not the problem. There is a top-level conference call virtually every day between the two services and regular contact between Bush and Blair. Nonetheless, the British were given a report that was close to the opposite of what Murray had first reported.

As fall changed to winter in 2002, Murray tried to continue to work Sabri through the Lebanese intermediary and kept pushing the Iraqi on the issue of defection. All this came to naught. The reports that Murray submitted reaffirming Sabri's intelligence were met with silence from the White House. During this time, Sabri warmed to the idea of defecting and stipulated, mainly, that his family would have to be evacuated with him; otherwise they'd be killed. This proffer was never pursued. In January, Murray discovered that the Lebanese intermediary had made off with Sabri's $200,000 and he tried to get another requisition for the Iraqi. He was refused.

Sabri's intelligence was buried, never conveyed to the Pentagon or to Colin Powell at the State Department. As the Niger claim made its way into the State of the Union Address and Powell prepared his UN presentation, Sabri's offering was nowhere to be seen.

I'm sitting with a man in a Washington restaurant. He's an official, highly placed for some time in the American intelligence community. I've known him for a while, and we talk, as usual, about many things.

That's the way it is with sources if you do investigative work. You build relationships; and those relationships matter. Because, over time, the best sources end up telling you the truth whenever possible. An outright lie, or an improper shading, might soon enough be discovered by the reporter. The source, generally, will be asked to explain. And that's a hard moment, one that can ruin this sort of transactional understanding.

This source has been right about many things. Always right, in fact. But we don't talk about this or that disclosure. It tends not to work that way. You talk around things, like walking around a tree waiting for a fruit to ripen and fall.

Today the discussion is about all the things that people don't know and the consequences of so much secrecy, on such important matters, in a democracy. You opine with sources about those sweeping issues; many reporters do, because that's the context in which we live—sources, reporters, everyone—and context matters.

Eventually, I segue to my riff about truth, about its being what works "in every relationship that you have that *does* work—between loved ones, a parent and child, friends, colleagues, employees and their bosses, citizens and their government." I've done this riff a lot. In this case, I add a caveat, that "truth is messy, uncontrollable, and it can be disruptive. But at day's end, it's all we have. And we—all of us—need to trust it. We need to start to trust truth again."

Then we sit for a while. He seems to want to tell me something, and I want to give him a moment to decide whether he will.

And then he edges in, ankle deep.

"We knew," he says.

"Knew what?"

"That there were no weapons in Iraq."

"Sure," I say, "people suspected. Define *knew.*"

He pauses. The waitress refills our cups. He takes a sip of the freshly poured coffee and waits until she's out of earshot. "Well, there was an amazing intelligence mission conducted before the invasion. A British intelligence agent, a real star, their best guy, met with a head of Iraqi intelligence at a secret location. Here the whole world is on edge, and they manage a secret meeting. And the top Iraqi guy explained it all: why there were no weapons, what happened to them, why Saddam was playing this strange game of ducking and winking. Pretty much everything that eventually came out. And it all made sense."

"How far ahead was it?"

"A few months, from what I've heard. Plenty of time to call off an invasion. Hell, you could call it off a week ahead if you really wanted to."

I sit thinking for a minute, jot down some notes. I mention Sabri. My source knows all about the case. "No, no. Much bigger and more complete than that, more detailed." He explains, accurately, that "Sabri was a diplomat, and we could never meet with him—we just passed him questions through an intermediary and waited for answers. No, this was the real McCoy. This was the Iraqi intelligence chief. He knew all there was to know."

I ask if the intelligence was passed to CIA and the White House.

"Of course. Passed instantly, at the very highest levels."

"And what did we say," I ask. "Or, I guess, what did Bush say?"

"He said, Fuck it. We're going in."

I ask the source how this might be reported, how closely held it was. He said only a "handful of people" know. Even he, with highest-level security clearance, knows only the barest outlines. He thinks the intelligence chief is the guy who "was just hung in Iraq, the guy whose head came off." He can't remember the name.

What about the British agent?

He pauses, considering his next step. "His name is Michael Shipster. Been around for a long time. Debonair guy. Brilliant. He's the best they have. But good luck finding him."

Chapter 3

Spies, Scones, and
the Werewolves of London

During what's called its Imperial Century, from 1815 until 1914, the British Empire was the largest in history, holding sway over a quarter of the world's population and a similar share of its land, stretching across more territory—from Canada to Africa, Asia to Australia—on which darkness fell at a given moment on any night. Its legacy in spreading concepts of law and government, economic systems, educational ideas, and everything from soccer to the English language is vast beyond reckoning.

When an empire falls, when it can no longer exert claim or control, it simply becomes a crossroads, where disparate peoples—the colonists and the colonized, rulers and ruled—go about their business, side by side, in a lingering aftermath.

That's Britain generally and very much London—a crossroads, from which paths leading in every direction are visible and tremors from all the planet's shifting plates, west to east, can be measured. There are few better places to see, and feel, the world as it is.

Late on July 9, the streets in central London are quiet for the most part, the last shops closing down and the city drifting off to sleep. Most Londoners won't know until morning that four thousand miles away in Islamabad, Pakistan's security forces are up early preparing to storm Lal Masjid, the Red Mosque. After an eight-day siege, it's in these predawn hours that Musharraf's forces move to break the stalemate that's settled over the mosque where militant religious extremists are holed up with automatic weapons and a couple dozen child hostages.

The mosque crisis has utterly consumed Pakistan for the past week and, as well, wide swaths of Great Britain, which is home to nearly a million people of Pakistani origin. The connection to Pakistan is like thick power cable—four hundred thousand Britons traveling to Pakistan each year—that runs a charge right through London. The situation is similar for the large and active UK communities from Egypt, India, and Saudi Arabia, of course, and a collection of African countries, including troubled ones, such as Sudan and Somalia, that the British once ruled.

By morning, as reports about the storming of the Red Mosque envelop the news, street protests and press conferences are launched by the various Pakistani factions around London. Clerics speak out. Police go on alert. Extra manpower is applied to the "ring of steel," a surveillance system of thousands of cameras that record every movement made in the old city.

Just another day in London. The city, after all, has been exploding with a kind of measured regularity in the past few years. The latest incident came less than two weeks ago. On June 29, two cars packed with explosive gas and nails were found in central London, ready to be detonated. The foiled attack brought to mind the devastating subway bombings of 2005, and when a Jeep filled with propane crashed into the main terminal of Glasgow Airport on June 30, it was more than enough to bump the terror threat level to "critical" and leave the country bracing for the next strike. Most troubling was the fact that those apprehended were not just educated and upwardly mobile—which many violent jihadists are—they were doctors. What can be done, the British press has fretted, when doctors take a second oath to do grievous harm?

Dame Eliza Manningham-Buller, the recently retired head of Britain's domestic intelligence branch,
MI5, added fuel to this fire with an article, just published, describing the threat to Britain by terrorists as of
"unprecedented scale, ambition and ruthlessness." She
pegged the number of terror plots in the works around
the country at thirty and the number of terror networks
at two hundred, with an aggregate membership of well
over fifteen hundred individuals. "It remains a very real
possibility," she warned, "that they may, sometime,
somewhere attempt a chemical, biological, radiological
or even nuclear attack."

Harrowing words, and intended to be. But sweet are
the uses of such adversity, the British insist, to focus the
mind on solutions, even ones that might be exported to
their American cousins. With its large Muslim communities and long history as a refuge—home to the first
legal protections of the rights of men and free speech—
Britain feels, with some justification, well positioned to
deepen its expertise in the hearts-and-minds struggle.

It's the Brits' term, after all. It was Sir Gerald Templer, the British high commissioner of Malaya in the
1950s, who first and famously uttered it. In those days,
as the British and their loyalist Malay subjects were
battling a communist insurgency, Templer remarked
that "the answer lies not in pouring more troops into

the jungle, but in the hearts and minds of the people." His initiatives—offering citizenship to disenfranchised groups, building villages for relocation, rewarding rebels who surrendered—have become standards for counterinsurgency campaigns since.

Which, incidentally, is the way the British now see the current landscape: as an insurgency of global reach—driven by Muslim radicals but not exclusive to them—that cannot be defeated primarily by force. By the early spring of 2007, the British government has stopped using the term "war on terror."

A Gordon Brown speech in late June, just days before his taking office, spelled out Britain's new path, how the effort "to isolate and defeat terrorist extremism" is "a struggle of ideas and ideals that in the coming years will be waged and won for hearts and minds here at home and round the world." To an especially large degree, for this colonial power turned crossroads, the fates of "home" and "the world" are one and the same.

That was a combination, of course, the British relied on in their glory, when their rather modest population controlled so much of the world. The Royal Navy, economic pressure, and crack British troops were not enough. The key was understanding acutely the peoples they ruled, getting enough of them invested in the British system of life and law to preserve order, and developing the "good offices" and loyal sources of a top-flight espionage service.

It's an hour's trip by train from London to Cambridge, a city of one hundred thousand happily dominated by the famous university, with its twenty thousand students and eight centuries of rich history.

This is where Sir Richard Dearlove, a Cambridge graduate himself, landed after he left his post as head of Britain's Secret Intelligence Service in the summer of 2004. He became master of Pembroke College, one of the oldest of the thirty-two colleges that make up Cambridge, founded in 1347 and boasting such graduates as William Pitt the Younger, Abba Eban, and Monty Python's Eric Idle.

Over the centuries, parts of the college were rebuilt, but the main arch is the original weathered sandstone. It opens onto a cobbled path that leads to a low doorway, and from there it's up one long flight of sloping stairs to Dearlove's office.

This seems like a particularly good place to keep secrets, locked in sandstone. When Dearlove's appointment was announced, Cambridge was quite clear that he wouldn't be doing interviews, and this came as little surprise. Controversies over Britain's role in supporting and joining America's mission in Iraq were beginning to surface regularly by then. Dearlove knew they were bound to get worse. As head of MI6, he was always the invisible man at the table, right next to Tony Blair.

He was at Cambridge less than a year when, in May 2005, the "Downing Street Memo" was reported in *The Sunday Times*. The leaked memo from July 2002 detailed a briefing to Blair by a British intelligence official, just back from America, simply known as "C." Anyone who's seen James Bond talk to "M"—the fictional head of MI6 and Bond's boss—is familiar with the single-letter designation for the head of British intelligence. The 2002 memo read: *C reported on his recent talks in Washington. There was a perceptible shift in attitude. Military action was now seen as inevitable. Bush wanted to remove Saddam, through military action, justified by the conjunction of terrorism and WMD. But the intelligence and facts were being fixed around the policy. The NSC had no patience with the UN route, and no enthusiasm for publishing material on the Iraqi regime's record. There was little discussion in Washington of the aftermath after military action.*

Those seven sentences, terse and pointed, caused an understandable fuss in the UK, then in the midst of an election.

But there wasn't much more revealed on the matter. Britain has developed a body of secrecy laws since the

296 • RON SUSKIND

passage in 1911 of the Official Secrets Act, which strictly prohibits the disclosure of "security and intelligence information." An amendment to the act in 1989 removed its "public interest defense" provision, which exempted disclosures of compelling public interest—the legal rationale that underlies "whistleblower" protections in the United States. On top of that, the intelligence services in the UK are not monitored by Parliament; there are no oversight committees as there are in the United States. The services simply report to the prime minister. Secrets thus stay secret, allowing known intelligence chiefs such as Dearlove a means of retreat. They can spend their retirements smiling coyly.

One of Dearlove's few revealing public appearances occurred, predictably, in America—a much less legally rigid terrain than the UK—and he took the opportunity to mention that the leaked Downing Street Memo was a draft, altered in some invisible way prior to delivery, and that if people wanted to know more they should check the "Pembroke College archives in 100 years."

But on one point in the conversation—a public discussion at the Aspen Institute moderated by James Fallows—Dearlove became animated and uncharacteristically revealing. "In any campaign, a leader, a general adjusts their strategy and tactics to the situation," he said, "and I think now what is vitally important for all of us in confronting terror is that we attempt to climb on to the moral high ground. For a variety of reasons we are not on it at the moment, and it's quite clear, if you analyze the chronology of the cold war, one of the reasons, I think, we ended up in such a successful position, because the West, unequivocally, at the end of the cold war, did occupy the moral high ground, and I think that if we are to be concerned with strategy and solving causes, rather than treating symptoms, a strategy that takes us on the moral high ground is absolutely essential."

He mentions the "moral high ground" three times in those two long sentences, and I'm hoping, in our discussion, that he'll see the connection between public honesty—between coming clean about a nation's true intentions—and the "climb" back up to that higher ground.

Because I know only a few key facts, and they might not be enough to compel him to offer the rest.

The secretary lets me into Dearlove's chamber, a rather unimpressive top-floor garret with a blond woodwork table and a sloping ceiling.

I've never met him before, and we talk for a bit by way of introduction. He's affable, but that's to be expected. He's a skilled intelligence officer who served stints in Nairobi, Geneva, Prague, and Paris before becoming head of MI6's most important office, Washington, in 1991. It was a post to which he brought particular affinities, having spent years as a boy at the Kent School in Connecticut. As an adult, Dearlove became a member of Kent's school board and traveled to America frequently for meetings—"something Sir Richard always used as a convenient cover," an old hand in America's clandestine service once told me.

The comment was only half in jest—various operational veterans at CIA have talked of how artfully Dearlove handled Tenet over the five years their jobs as intelligence chiefs overlapped. "He was George's case officer," is the refrain—one that gets to the heart of the American problem of placing politicians or generals in the role of intelligence director, only to see them outmatched when they have to face off with the heads of other services who've spent decades in operations.

I lay out what I know, which is just the outlines—that there was a mission in the months before the war in which Michael Shipster met with an Iraqi intelligence chief. It's clear some of my facts are wrong. My source in Washington, knowing only the barest details himself, thought the Iraqi was the guy whose head came off when he was hanged in January. Barzan Ibrahim al-Tikriti was his name, a former director of Iraqi intelligence and an adviser to Saddam, but not the official head of the service at the time the regime fell. Tikriti, a thug who'd killed thousands for Saddam since the 1970s, suffered the fate of a too-long rope—meaning it was too long a drop from the gallows before the noose snapped taut—and the indignity of sudden decapitation.

"No, no, it's not Tikriti. He wasn't the guy," Dearlove says after a bit.

Rather than playing the name game with an org chart of Iraqi intelligence, we move along, identifying the Iraqi as simply "an intelligence chief." Soon, the rest falls into place. Yes, Shipster had a secret meeting in the months leading up to the war, arranged through an intermediary.

Dearlove has trouble nailing down the exact date of the meeting and hints that there may have been several. He's sure, though, that one was in February.

I run through more particulars, as I understand them: that the intelligence chief told Shipster there were no WMD, described why that was case, and went on to explain how Saddam was worried that his neighbors, especially Iran, would discover he had none of the weapons they most feared.

Dearlove confirms all this. Then cuts me off. "How do *you* know about Shipster's visit?" Only very few people, he mumbles, on either side of the Atlantic, know any of this. He removes his glasses, rubs them with a handkerchief, and then shakes his head. "Yes, it did happen."

I mention, fishing a bit, that I understand the meeting took place in Iraq, that Shipster sneaked in and out of the country, more or less undetected. Dearlove frowns. "It didn't happen in Iraq," he says. But he confirms that Shipster's precious haul of intelligence was passed immediately, by February, to Washington. Everyone "at the top" knew all about it—he and Tenet, Blair, Bush, and Cheney.

Dearlove, of course, knows how this takes a sledgehammer to the case for war made by both Bush and Blair and digresses to do some tidying up before the hammer falls. There's something he's been wanting to clarify. The famous sixteen words from the State of the Union—"The British government has learned that Saddam Hussein recently sought significant quantities of uranium from Africa"—is a misstatement. "The UK's interpretation was that Iraq was making *preparations* to *attempt* to buy" the yellowcake, he says. Bush was suggesting Iraq "had reestablished its program" to create nuclear weapons. "We disagree on this issue."

But a larger question is now draped around us both. What happened? Why didn't Bush—or Blair, for that matter—act on the startling intelligence from Ship-ster?

Dearlove talks instead about intent, about how the daring mission was an eleventh-hour "attempt to try, as it were, I'd say, to diffuse the whole situation. . . . I mean, it was our willingness to put someone in a pretty exposed position, which is a tough, tough thing to do."

Yes, and the mission succeeded. "Yeah, but you know, we're good at these things. That's what we're paid to do!" he says, glowingly, about one of the era's great espionage coups. "You know, we're rather better at it than the Americans."

But it seems the Americans were unappreciative of the gift, which arrived in plenty of time to stop, or certainly delay, the invasion.

Dearlove nods—yes, a month or so before the March 19 invasion is certainly enough time to reconsider matters. Which brings him to a fault line, a place he didn't expect to be a few minutes before, or maybe ever. Some secrets stay secret, and a few people he knows well probably hoped this would be one of them, a disclosure that refutes countless public statements by duly elected leaders, both American and British, about the serious matter of war. Dearlove is classically educated and well read, a student of history who understands that even tyrants, in dark ages, thought long and hard before committing their men to war, and—as a general rule—were careful about what they cited as the "just cause." It could be conquest of a hated enemy, the return of lands to their rightful owner, an insult that soiled a nation's honor and had to be avenged, a holy cause of God. But the men fighting and dying deserved to know why. Lie to them, and those soldiers might well turn on the castle, with a few ambitious generals in the lead.

He looks out the window for a moment, across Pembroke's Main Green and toward the college's majestic Old Church, built in 1355 with funds from a papal bull—a decree in the form of a letter signed by and carrying the authority of the Pope, who then ruled a vast kingdom built on faith. The British would eventually turn the great church into a library, befitting an empire built on reason and the concepts of justice that flowed from it. Now it's the turn of their successor, Promethean America, to make choices only an empire—with it awesome and dangerous latitude—can make.

In this case, they seemed to have decided to lie to the world. But why?

"The problem," Dearlove says, finally, "was the Cheney crowd was in too much of a hurry, really. Bush never resisted them quite strongly enough." His voice trails off as he looks beyond the Old Church to the temples of Washington.

"Yes, it was probably too late, I imagine, for Cheney," he says, about stopping the invasion. "I'm not sure it was too late for Bush."

Dearlove then turns and stares at me over the rims of his glasses, an intelligence man for thirty years whose greatest triumph was ignored.

He has finally arrived at the question of culpability, of moral judgment.

And he wants to be very clear.

"I don't think it was too late for Bush."

On the train back to London, I review some materials I've gathered on Iraq's intelligence operation. If it's not Tikriti who met with Shipster, then who?

I call my assistant in America and have him run through my file on the Iraqi leadership. The head of Iraqi intelligence, or IIS, in the last years of Saddam's regime was a man named Tahir Jalil Habbush. Sometimes he added "al-Tikriti" at the end of his name— meaning "from Tikrit," the city in Iraq from which Saddam and many of his senior lieutenants hailed—and that may have accounted for the confusion of my source in Washington. Habbush had not been seen since Baghdad fell. He'd been placed on various Most Wanted lists, including one of the Chief Criminal Court of Iraq. They posted a $1 million reward for his apprehension in 2005. Nothing has come of it. Most experts believe that Habbush, like several other members of Saddam's inner circle, is dead.

The reason I need to know all this so quickly is there's a bit of fortuitous scheduling at play. I'm rushing back to London to meet with Dearlove's number two at MI6, the service's assistant director until 2006, Nigel Inkster.

We meet for drinks in the midafternoon in the lounge of a London hotel, not far from where Inkster now works at one of the city's major think tanks.

I tell him I've just met with Dearlove and that he and I talked about the Shipster mission. When I begin discussing it with Inkster, I mention Habbush as the Iraqi intelligence chief, and he confirms that, in fact, Habbush was Shipster's secret contact. "Though," he adds, "I'm not sure how Habbush managed to get out of Baghdad for the meetings."

During his thirty-one years in Britain's Secret Intelligence Service, Inkster was stationed across three continents, worked on all manner of "transnational issues"—which means various types of cross-border trafficking and terrorism—and spent his last two years, from 2004 to 2006, as assistant chief of MI6 and director for operations and intelligence. He's a natty fellow—a Chinese speaker with a degree in Oriental Studies from Oxford—who carries a passing resemblance to Jeremy Irons, pinched and world wise, with delicate bags beneath his eyes.

Inkster orders a vodka, and we talk of espionage. The whole Shipster business was very tightly held on the British side. "It was Richard's thing," he says. "At SIS, there's this very strict need-to-know culture."

But the yield, Inkster admits, was particularly illuminating. Habbush described to Shipster "Saddam's focus on his own image and his regional enemies. That was key." Hussein was concerned about Iran, other neighboring competitors, and dissident groups inside Iraq. Inkster recounts that Saddam didn't think a U.S. invasion "was a serious proposition," that the United States would be so foolish. "He couldn't believe it was going to happen. I think that was the fundamental problem with the whole thing—we didn't read ourselves into Saddam's mind adequately, and he was completely incapable of reading himself into ours. And we missed the thing altogether."

But, of course, we did, with the help of Habbush and Shipster, read ourselves into Saddam's mind. Ink-

ster agrees that this was the idea that drove the Shipster mission. It was an attempt to "get inside Saddam's
head." It's what Blair wanted.

"You have to bear in mind that at that point the UK
and the U.S. were in very different positions. I think
within the USA there was widespread resignation that
this was going to go ahead . . . and it was already more
about preparing for the aftermath. Whereas within
the UK, of course, the whole thing went right down
to the wire. And everyone was trying to find a way
out, if that could be done. Why exactly did we explore
this possibility?" he asks—referring to the Shipster-
Habbush encounters—before answering his own question. "Because I think everyone—Tony Blair, as prime
minister, knew that he was going to struggle to take
his party and the country with him" into Iraq. So Blair
turned to Britain's crown jewel—its seasoned intelligence service—to do something, Inkster says, that
would alter "the course of events."

And when this prized intelligence was ignored by
the United States?

"You know, the feeling was that this was a decision
the U.S. had made way back and, you know, that was
the defining perception."

He thinks it all over for a moment. The lounge is
now bustling with the after-work crowd. Life goes on.

There's been a clamor steadily rising for years, for British troops to come home, that Iraq—"Blair's folly"—was a terrible error that deepened year by year, making this secret, last-chance mission burn ever more brightly, for the few in the know, as Britain's what-might-have-been moment.

With Blair on down, Inkster says, there were "a lot of different emotions" about what they had discovered, and about America's dismissal. But the view, at day's end, was that the United States was "like a runaway train. There was nothing that was going to stop this."

I wander London that evening thinking of Dearlove's statements about how important it is to regain the moral high ground, how the tactical battle, day to day, year to year, is unwinnable. The latter is something almost everyone quietly agrees about, from Rolf to Gordon Brown and Dame Eliza. But restoring moral energy is maddeningly difficult, especially on a terrain of constant confrontation and accusation, where the expression of doubt or the admission of error is so easy to cast as weakness. Wrongly cast, on balance. The courage to be honest about past mistakes and lessons learned is an emblem of the secure and the forthright, of people and nations that win respect over time and earn the chance to reach for those moral uplands. But the game played everywhere is to admit nothing, stay firm, and hold that line. Words are such convenient weapons that they're difficult—as difficult, in a way, as nuclear weapons—to lay down.

In my hotel room that night, former Pakistani prime minister Benazir Bhutto is on the television, being interviewed about the crisis at the Red Mosque. Bhutto is easy to find—she has a penthouse apartment near Regent's Park—and as the mosque crisis intensifies, she's talking nonstop to the British and international media. It's now public that she's been in conversations since the past fall with the Musharraf government about her return to Pakistan and a possible power-sharing arrangement—a deal encouraged by the United States.

"The Red Mosque symbolizes the failure of the policies of appeasement," she says, coming on like a warrior queen. "Every time the militants are in a corner, instead of prosecuting them or arresting them, there's a cease-fire or a peace treaty or an amnesty. They get out of the corner, they're more emboldened, and they come back with a vengeance."

Appeasement? More than fifty militants are already dead as the Pakistani army continues its assault on the mosque compound, home to three hundred women and children in a middle-class section of Islamabad.

But Bhutto is just singing the full-throated stanza

of a common refrain: that force works, in both word and deed. The United States wrote the latest version following 9/11 to challenge a new, hydra-headed foe— Bush's "with us" mantra at heart a call for states to use force against the tactic of terrorism. Six years along, it's clearly showing its inadequacy, and nations, with Britain in the lead, are struggling to find replacement strategies.

My breakfast companion at the hotel the next morning, David Omand, is the leading authority on one strategy: techno-control. His great innovation, as Britain's terrorism chief from 2002 to 2005, was the installation of cameras—thousands of cameras—to place almost every square meter of central London under video surveillance. The grid is too vast for much in the way of active monitoring—though techniques are being developed to identify problems in real time with computer imaging. The plan has, however, provided law enforcement with after-the-act images of terrorists on the move, which have led to the identification of suicide bombers and in some cases their accomplices—photos that viewers now regularly see on television. The loss of privacy for Londoners was, at first, controversial, but, Omand says with satisfaction over his eggs, "they've become surprisingly comfortable" with the surveillance "in a rather short time.

"People go about their lives; this is just an invisible part of that life."

Omand, now out of government, has been making the rounds in the past year giving frank and elegant speeches from Lisbon to Palo Alto that are regularly cited for their straight-talk renderings of the arc of the counterterrorism battle.

He, like others, often notes that "preemptive secret intelligence is the key," but Omand has been brutally frank of late about why it has become so problematic to gather: "It is probably the case that by using methods such as extraordinary rendition, deep interrogation, indefinite detention, and targeted killing, that the U.S. has lost more—and her allies with her—than we have gained in short-term relief from terrorist attacks," he told a packed house at Stanford University in February. Omand's closing remarks at the Stanford speech have been collected and pinned to the wall—or rather the My Documents file—of countless counterterrorism officials worldwide, a reminder of why extraordinary extrajudicial means play into the hands of the opponent:

"The aim of the al Qaeda leadership for the present phase of their campaign is not just to attack us. It is to try to create the impression throughout the Muslim world that a global struggle against oppression is under way in which violent jihad against us is a personal duty since, in their eyes, the policies of the US and its allies towards the Muslim world are incurably discriminatory and at heart colonial. Through constantly tempting us into over-reaction, they want to expose our values as fragile and hypocritical, suppressing civil rights at home and supporting apostate and repressive governments overseas. We should recognise their motive as the well-understood tactic of the revolutionary through the ages, and not fall for it.

"We also," he concluded, "have to develop our own convincing narrative to counter domestic radicalization, of the better alternative our society offers. As the Director General of the UK Security Service has recently reminded us, 'if the opinion polls conducted in the UK since July 2005 are only broadly accurate, over 100,000 of our citizens consider that the July 2005 attacks in London were justified.' "

Omand says over breakfast that, yes, privacy will be further compromised—with more electronic surveillance measures instituted "in response to future attacks"—demanding "even more transparency and safeguards as to oversight." In that progression, he says, governments can't "stay where they are—they'll need to embrace ever-increasing standards of ethics and accountability. That's the challenge—a real tough one."

The first part of the equation, however, is already

under way. Cities in Europe and across America, starting with Washington, are signing on to high-budget
"ring-of-steel" surveillance systems, encouraged by
Omand's speeches and the marketing departments of
large security contractors.

It has become something of an industry, he agrees,
but so have a growing acceptance and awareness on the
part of the public about the complexities of the battle
and the fact "that a bomb can go off but the world
doesn't necessarily change."

The word he likes is *fortitude*—a British favorite—
that will settle into "a kind of bounce-back flexibility":
violent extremist acts are going to happen, but that
doesn't mean that the world is "some sort of polarized
battleground. . . . We'll keep right along working on
the important issues, the big ones, to lessen the world's
strife in ways that are visible and consistent."

And along the way, there will be lots of video cameras. As he sallies forth, we chat momentarily about the
hidden cash flow beneath his steel rings: tickets, a river
of them, for parking and traffic violations. In all, an
ideal governmental solution: invasion of privacy, rev-
enue neutral.

Back in my room, I call Tripoli.

That's where Omar Bakri, an outspoken cleric who's
been accused of issuing messages on behalf of al Qaeda,

now lives. He left the UK in August 2005 when it came out that the British government was considering treason charges against various radical Muslim clerics, and then was forbidden from returning.

I met Bakri in the lobby of this hotel two years ago, a few months before his ouster. At that point, he was a leader among the radical clerics in the UK, men who move around London, with their impressive hair and competing identities, like werewolves. By day, they give speeches, write online columns, and visit mosques, coming up just short of inciting specific violent acts—the legal threshold here. By night, they troll "safe house" apartments in the city's Muslim districts, offering fatwas, clerical interpretations of the Koran and the Sunnah, which can either bless attacks or stop them.

Bakri, a Syrian-born rabble-rouser who settled in London in the mid-'80s, is still the most widely known of the clerics. In 1998, when al Qaeda car bombs exploded in front of two embassies in East Africa—killing 212 people, 200 of them Kenyan, at the embassy in Nairobi—Bakri released prepared statements from bin Laden. Roland Jacquard, a French terrorist expert and regular adviser to a number of governments, said that "every al Qaeda operative recently arrested or identified in Europe had come into contact with Bakri at some time or other."

But there's a hidden dimension to it all, which I

glimpsed when we chatted in the hotel lobby two years back. A British intelligence official told me that Bakri had helped MI5 on several of its investigations. After Bakri finished one of his long explanations of how the Koran allowed him to defend himself "but not to attack nonbelievers just for being nonbelievers," I mentioned his secret assistance to police. He became flustered. "I'm upset you know this," he said grimly, as one of his deputies looked on from a distance. Of course, such a disclosure would have undermined his credibility among young radical jihadists, his constituency. I asked, then why do it, why help the police? He paused. "Because I like it here," he said. "My family's here. I like the health benefits."

A few months later, the July 7, 2005, bombings killed fifty-two and injured seven hundred—London's worst terrorist attack—and all such backdoor arrangements were off. A new terrorism law was passed, speech was further constrained, clerics were rounded up, and Bakri was out.

On the phone now he says, "I'm doing some important work" in Libya. But he also mentions that his health isn't terrific, and he clearly misses his family— he brings them up several times in the short call—and his old role. He says the British government, "whether they admit it or not," misses him, too. "We were able to control the Muslim youth," he says on the grainy mobile phone connection from Tripoli. "The radical preacher that allows a venting of a point of view is preventing violence. Now many of us are gone or in jail, and we've been replaced by radical jihadis, who take the youth underground. You don't see them until the day they vent with the bombs."

His statement is self-serving but not far off the mark. In fact, Britain's dance with its radical clerics is a revealing illustration of the elasticity of self-interest, how it can stretch across seemingly insurmountable divides. The demonstrations that Bakri and other clerics staged clearly roused angry Muslims—and may have somewhat increased their numbers—but they also drew them into daylight, into the public space, and gave them a voice. If someone, anyone, responds to that voice, it is, by definition, dialogue. Bakri enjoyed his notoriety and was willing to pay for it with information he passed to the police. He knows his London followers are staging a protest this morning about the Red Mosque, and he offers a halfhearted telephonic outburst: "Yes, because all Muslim brothers and sisters must oppose the tyranny of Musharraf, like the tyranny of Bush and Blair!" It's clear he would trade almost anything to be back, leading them—anything—at a time when the British authorities, reeling from the latest attempted bombings, happen to have great needs. It's a fabric of subtle interlocking needs: the Brits need be in a backchannel conversation with someone working the steam valve of Muslim anger; Bakri needs health insurance.

At lunchtime, demonstrators gather in West London near the Pakistani embassy. There are fifty of them, men in *kameezes*, chatting, and women in *abayas*, lined up with signs, some with their eyes visible, some not. Only the men speak. Many are young and filled with fire; everyone is prepared to be a spokesman. They crowd around the few reporters milling about.

I spot some werewolves in the crowd, including one prominent one, Yassir al-Sirri, the Egyptian exile who has been accused of setting up Ahmed Massoud, the tough, charismatic former leader of Afghanistan's Northern Alliance. Two days before 9/11, Massoud was approached by a pair of Belgian cameramen, who said they were doing a documentary and carried a letter of introduction from the Islamic Institute of London, al-Sirri's organization. Massoud received them. Their camera was in fact a bomb. Massoud—the natural leader of any American response to the Trade Center bombings, just days away—was now dead. When I met al-Sirri in 2005, at the Wembley Hilton, he said, with brio, that the letter was a fraud, that his organization's "logo is a rising sun and in this letter was more of a mid-morning sun."

Egyptian president Mubarak has been pressing the British to extradite al-Sirri; he faces two death sentences in Egypt, and he's become careful. "I'm more cautious now than I was," he says through an interpreter. That caution has turned him into a man of talk rather than action. As Wendy Chamberlin says, "people just want to be part of the discussion," and that's what al-Sirri has become: just part of the dialogue. He opines that Salman Rushdie's imminent knighting by the queen—an act that drew angry words in an audiotape released yesterday from Zawahiri—"just plays into the hands of people who are terrorists, gives them a way to justify their actions. We want to see an end to the atrocities that happen all over the world."

Behind us the noise is rising. Anjem Choudary—Omar Bakri's longtime deputy and now his replacement as London's most public radical cleric—takes the megaphone and lobs invective at unseen enemies: "Musharraf is a terrorist! . . . Gordon Brown is a terrorist! . . . We stand with our brothers and sisters in Pakistan!" Each cry is repeated by the crowd with the call-and-response rhythm familiar to anyone who's been to a Southern Baptist church.

After ten minutes, Choudary hands off the megaphone and steps back for a moment's rest. "We've been humbly blessed by Allah with media," he says, mopping his brow. "That is for the good. Even if the crowd isn't as large as some, our brothers in Pakistan maybe will hear us. They'll know we are with them."

The "we," in this case, are a diverse lot. But that's not uncommon: "true believer" is not a résumé-driven avocation. One man in the sidewalk battalion is a retired engineer. Another works at an employment agency. A few are middle-aged, through most are under thirty. One nineteen-year-old kid from Bangladesh says he wandered around London for a year before "Islam gave me a home." An older brother-in-arms, twenty-seven and sporting a long beard, says the faith makes "me feel free, free from what has always confused me as some chap who just worked to get by, and please my parents, and find a nice girl. Now I follow the Koran. And if I follow it precisely, as best I can, I can't be wrong." Whether any of them is following the Koran the right way or the wrong way—being true to Muhammad's vision or subverting it—is almost beside the point to most of this assemblage. This is about how they *feel*. Unless, or until, they can feel this sense of well-being and belonging somewhere else, no one's going anywhere. They tend not to be drawn here by a predisposition to violence, experts on jihadist recruitment tend to agree.

These experts, scholars who've interviewed hundreds of radical jihadists, also agree about what provides the impetus to often destructive action: loyalty to one's brothers and sisters in the fight. The same goes for the infantry company in a war. Just ask any Medal of Honor winner; they'll all say the same thing: "I did it for my guys." The question is what will define their time on jihad's street corner.

One thing is indisputable: this group of fifty likes the way it feels to be *in the discussion*. And that goes from top to bottom. Bakri, in Tripoli, would give anything— including signing on as a regular police informant—to get back into the discussion. None of them—the engineer, the headhunter, the kid from Bangladesh—can stop talking: about the Koran, about the dangers of desire, about their parents' values; and they don't seem ever to want to stop. Despite their ninth-century attire, this makes them all very modern, denizens of this era of connectivity and instant images and communication, where representative individuals often feel like, and at times become, actors in vast global dramas.

Choudary, leading them through this noisy valley, is of Pakistani descent, but British born. He's an attorney, a wide-set, nice-looking chap in his late thirties who once practiced law not far from this fashionable neighborhood in West London. He says he knows that his rhetoric "approaches the line" of incitement. But "the government knows, or should know, that we're not planning anything here; it's just talk." In a precise echo of Omand's comments—about how radical Muslims are "constantly tempting" authorities to "expose our values as fragile and hypocritical"—Choudary adds, "we're testing whether the British government is really true to the idea of freedom of speech and innocent until proven guilty. Let's see."

Then, with a nod, he steps forward again to take the bullhorn.

Out of earshot in her penthouse apartment near the Royal Albert Hall, Benazir Bhutto is having a tea party.

Or she will be in a few minutes, once her long and sumptuous table is set.

If Choudary and his jihadist kin prowl the lowlands, erupting with self-conscious outrage, Bhutto inhabits the mountain peak, the cloud city.

Today she's puttering around the sprawling apartment, receiving visitors in her bare feet. The line is long. That's because it's now common knowledge that Bhutto's moment—her latest moment—has arrived. She is at the center of a peak-to-peak gamble between the United States and the world's most dangerous country,

Pakistan—another test of the current reach of American power.

In the spring of 2006, Bhutto's representatives approached the State Department with an idea about Bhutto possibly returning to Pakistan after seven years of self-exile. But it wasn't until widespread demonstrations the following spring, after Musharraf sacked the country's chief justice, that the White House began to seriously entertain Bhutto's proposal as a way to shore up an embattled Musharraf. The Pakistani strongman might be called a single-issue ally. All the United States really cares about is how well he's using his army and intelligence service to police the de facto capital of violent extremism that has emerged along his border. This would include al Qaeda, now fully reconstituted and growing in the Federally Administered Tribal Areas, or FATA, a historically lawless region in northern Pakistan; and the Taliban, which moves its forces freely across the meaningless mountain borders between Afghanistan and Pakistan and runs its operations from Quetta, a provincial capital in western Pakistan. Among the added layers of complexity is the fact that both groups have long been supported—were even assisted in their founding—by Islamists inside the same Pakistani military and intelligence services called upon now to lead the counterterrorism fight.

From every angle, things have not been going well for Musharraf in this battle—not for years now. One reason may be that he's a dictator with declining support in his country. Each year, he seems to be backing further away from promises he made to the United States and his own people in the aftermath of his 1999 coup, promises to restore democracy to Pakistan. Not that democracy and Pakistan have ever been a very good fit. The country has had only eleven years of elected government since Bhutto's charismatic father, Zulfikar Ali Bhutto, was overthrown in 1977 by General Muhammad Zia-ul-Haq. Zia, an unlikely combination of military leader and fervent fundamentalist, placed scores of bearded fellow radicals in uniform. Many are now generals and colonels and intelligence chiefs, a group more religiously messianic, on balance, than the Pakistani population they serve, and theologically aligned with any number of radical mullahs inside the Taliban or taking up residence in the FATA. In short, it's an unholy mess, dominated across decades by dictatorial power wed, in an arranged marriage, to religious authority.

The United States, preaching since 9/11 about the power of democracy to break precisely this sort of union, has been looking ever more hypocritical in its might-makes-right support of Musharraf—precisely the kind of hypocrisy that corrupts America's moral authority. The United States reluctantly began to change course only when it became clear that Musharraf's might was significantly less than advertised.

Into this walks Bhutto—as complex a character as any on the world stage—who had her first formal meeting with Musharraf a few months back, in January in the UAE capital, Abu Dhabi. Musharraf was surprisingly friendly, Bhutto later recalled, and they talked through a long list of terms. Each knew the landscape well. Bhutto had twice been prime minister—not particularly effective either time—and both were clear on the idea that Musharraf would focus on directing the military and intelligence services in ways that would make the United States happier, while Bhutto would deal with long-neglected issues of social, economic, and diplomatic policy—her strong suits. But form, in this case, is tightly tied to substance. Bhutto, who managed from afar to keep her hold on the Pakistan Peoples Party— the party created by her father, and Pakistan's largest— would finally bring the seal of democracy back to her home country. Yes, the democratic ideal was faltering in Iraq, but it would triumph in Pakistan. That would happen, the United States felt, if the country could just ensure a reasonably fair election—all but guaranteeing Bhutto's electoral success—and get her and Musharraf to work out an arrangement for him to take off his uniform and retain the post of president, a kind of chairman of the board in the Pakistani constitutional system. Hence, the plan. Like most plans the United States has recommended, Musharraf seemed committed to carrying it forward when he was good and ready.

But in the late winter of 2007, Pakistan started to explode from within. Musharraf suspended the independent-minded chief justice of the country's Supreme Court, Iftikhar Chaudhry. The judge defied him—traveling the country behind swelling crowds—and soon became a vessel for Pakistan's growing impatience with dictatorship, especially among the country's burgeoning middle and professional classes. Black-suited lawyers, throwing Molotov cocktails at tanks, became the symbol du jour. Then Musharraf, already battered, woke a week ago to the Red Mosque crisis—a sign of things spinning even further out of control.

And the world's eyes, two by two, seemed to turn to Bhutto.

"Of course everyone is now very anxious to know how things are going with me and General Musharraf," Bhutto says, sinking into a vast maroon couch. "*So, Bibi, when will you be going back, and how will things proceed?* It's all turning into quite a test of principles—of democracy and its power. Truly amazing," she avers.

Bhutto—corrupted and corruptible, daughter to an executed leader, wife to a hustling playboy, mother to a bare few achievements, beyond her three children, after a quarter century in public life; proud, elegant, and vain Bibi, a woman who twice managed to lead one of the world's most paternalistic countries—the world's second largest Muslim nation—has become the unlikely vessel for the very concept of democracy.

Is she up to it? Her most faithful supporters—people who'd walk through fire for her—say maybe. They've seen what can happen when she's in a tight spot and is given a way out, how swiftly, how effortlessly, she can choose expediency over principle.

But after years in exile, living on the ill-gotten millions that she's told friends is the tribute collected, as a matter of course, by leaders in South Asia, she's trying to finally become the woman her father, in his final days, told her she was destined to be.

It's her last shot.

"The only way to attack terrorism is to be firm and consistent in support of democracy," she says, practicing her lines. "The people of Pakistan have tasted democracy, but only infrequently, and now they are hungry for more."

She works the oratory, as though she is speaking to an unseen audience, one that awaits her, talking in broad terms about needing "rule of law and not rule of men";

about how "people still admire America, as you can see any day by measuring the line at the visa office"; and how she "never lost faith in the wisdom of Pakistanis to determine their own paths."

The "real enemy, though, is Muslim radicalism," and that's "an opponent I know as well as anyone." With that, she finally finds her rhythm, quick and cutting, prosecutorial, running through the struggles of her own past, ticking off the names and ranks of enemies, starting with the men who hanged her father and progressing to present tense, listing men in Musharraf's government, right now, who are helping radicals inside Pakistan, from Karachi to Islamabad to the lawless tribal regions.

She can go on like this for hours, and speaks with special venom about someone she says is of immediate concern: Brigadier General Ejaz Shah, a Pakistani intelligence chief, formerly head of the country's Intelligence Bureau, who was once "the handler" of Osama bin Laden and Mullah Omar, and who gave refuge to Omar Sheikh, the man who killed Daniel Pearl, even after it was clear the latter had committed the murder and was being sought by Pakistani police. "Ejaz Shah kept Omar Sheikh at his house for a week," she says, "and briefed him about what he should and shouldn't say about the radicals inside Pakistan.

"This is tyranny, these men are tyrannical, and I must show Pakistan how democracy can really work for them, for the many poor people." The transition is strained; the choice between tyranny and democracy seems, in her mind, at best a hypothetical. When she mentions what democracy will provide in the way of health care, education, jobs programs, and direct government aid to the huge impoverished segments of Pakistan that form the seedbed of radicalism—most of the country, in fact—it's just talking points; the fire's gone. Bhutto, after all, is from a feudal family, one of a dozen landowning families that have ruled the country since its founding. She knows "the many poor" mostly as servants or as faces lining a parade route.

This is a plain fact, of course, for many rulers in her part of the world. They tend to come from the privileged pinnacles of profoundly depressed societies. In Pakistan, a country of 170 million—now larger than Russia—two-thirds of the population lives below the poverty line or just barely above it, and 50 percent of adults are illiterate. Readers or not, they've witnessed an ongoing, decade-after-decade struggle for raw power— a game that is distant from their lives but enthralling— with the Bhuttos as stars. As narratives go, there's none in the Muslim world to match that of the daughter of a slain leader who becomes prime minister, only to be compromised by a corrupt husband and her own appetites. She's loved, hated, envied, despised, but they can't take their eyes off her. And if her appreciation for the virtues of democracy is roughly equivalent to that of Boss Tweed or Richard Daley, that's understandable. She had to learn how to outplay the boys. The great question—and global fascination—is about where that story goes from here.

But the guests are arriving. In a flowing, funky *kameez* of bright orange and lime green, Bhutto pads across the rugs to greet them—Tricia, a socialite; Sana, a wealthy Pakistani friend; Ginny Dougary, a feature writer with *The London Times*; Victoria Schofield, an Oxford buddy who's stuck by Bibi for years while writing scholarly books about Pakistan; and then Victoria's husband, an investment banker, and others, taking chairs at a long table filled with scones—piles of scones—and cream, jam, pastries, and cucumber sandwiches. High tea is served as Alan Duncan, a Conservative member of Parliament, drops in.

Festiveness warms the room. The repartee is light and quick, about scandalous turns among the powerful and celebrated in London, as well as Musharraf's woes and Bibi's hippy-ish *kameez*.

Duncan, the first openly gay Conservative MP, is a

stitch, and he and Bibi trade tales of the days when he ran her successful campaign for president of the Oxford Union, helping her become the first Asian woman to win the honor.

"Maybe, Alan, you should run my campaign now."

"Too busy, my love," he says, though he's ready to set her up with British foreign minister David Milbank. "Not that you need me now, Benazir."

And then everyone is chatting about Bibi's great good fortune.

"If it gets any worse for Musharraf," one guest starts, "it will open the door even wider for Bibi," another finishes.

"Can you contest the election from here?" Alan queries, jumping into campaign manager mode.

"If I don't go," she says, "my party suffers."

"I suppose you can't get enough television coverage over there, broadcasting from here."

"No, no, I must go."

"If you want to land somewhere," he says, "would they arrest you on the spot or turn you back?"

"Depends on what I arrange ahead of time," she quips. "Which should I prefer?"

Talk of Pakistan now reminds Bibi of a ribald story she heard about Shaukat Aziz, Musharraf's current prime minister. "Shaukat tells everyone that 'there's no woman I can't pick up in two weeks.' But when he tried his charms on Condi Rice, oh my.

"She just stared him down," Bibi exults. "She *withered* him."

Victoria, the author, mentions the spread of world jihad and how terrorist training camps were recently discovered in the Lake District, Britain's mountainous northwest, where Londoners vacation. "It's positively surreal."

Everyone agrees, and discussion swirls about how so many of the UK radical jihadists are second generation. "Their parents came here and worked their little butts off to open shops and whatnot," says Ginny, the *London Times* writer. "Now the kids are saying, 'Mom, Dad, you wanted me to find meaning in my life. Well, now I've found it.'"

"It's like Pakistan has come to London and London to Pakistan," Alan exclaims, reaching for a scone. "Benazir can now be a bridge between all these communities, from Pakistan to London."

Certainly one can hope, sitting high above London. Compacts may be made, under duress, between governments. Power-sharing arrangements struck. Deals cut.

But all of those gathered here in the clouds for high

tea suspect something, a hard and disquieting thing that they'd rather not discuss: to have true clout in the modern world, you have to walk through the valley and survive it.

And you can't get there from here.

When David Omand, among the most astute professionals in counterterrorism, talks of creating a "convincing narrative" about "the better alternative our society offers," he's alluding to a large and diffuse weave of example and impression, action and word. But what he, the Brits, and White House officials hopeful about the prospect of Bhutto's return are up against is one of the strongest counter-narratives in recent history: the story of bin Laden.

Every human story has been told and retold countless times, but the story of the "wayward prince" has a special status, found in the fables of virtually every culture. Most commonly, it is a prince who leaves the castle, a corrupted place, takes a vow of poverty, and walks, plain-clothed, among the people. It's a story that proceeds along the classic heroic arc: a character who abandons the familiar, confronts dangers and challenges in a twisting journey, and returns home altered, with hard-earned insight. In some cases, the prince leads the people against a wicked king. In others, he eventually returns to the people to live out his life, preferring them despite their penury.

It is clear that bin Laden and his deputy, Zawahiri, understand the power and varied applications of their version of the fable: a wealthy son of the Saudi elite who left it all behind, embracing poverty to walk among the people and then rise up to challenge the "apostate" Arab regimes and their oil-addicted American sponsor. The appeal of this story in a world where half the population lives on less than two dollars a day is hard to overstate. This is a simple fact, clear to U.S. officials—especially at the State Department—who, for years, have been quietly conducting surveys about the world's impressions of key figures and nations.

During the summer of 2007, a large survey of Pakistani opinions was being conducted by a private firm in Northern Virginia that matched results with those that internal U.S. polls had been collecting. The survey, of 1,044 Pakistanis across 105 rural and urban "sampling points," showed bin Laden with a 46 percent approval rating, Musharraf with 38 percent, and George Bush at 9 percent. Sixty-six percent believed America was acting against Islam or was anti-Muslim, and only 19 percent had a favorable view of the United States. It is worth noting that a similar survey conducted in October 2005—after U.S.-led efforts to help earthquake victims in Kashmir—showed a U.S. approval rating of 46 percent.

Polls are ephemeral, snapshots whose relevance quickly fades. But the battle of competing narratives, and competing strategies, is central to the decisions of countless young Muslims, mostly men, who inhabit London and Karachi, Riyadh and Kabul. And also New York.

But, as in other areas, London—with its combination of legal protections and jihadist fervor—provides unique avenues for understanding what guides combatants of all shades, on all sides.

In a northwest suburb of the city is a two-story house with a nondescript white façade, two doors leading into a residence and an office, where you can find a man wanted by the United Nations, the Saudi government, and the U.S. Treasury Department. He's a doctor, a surgeon from Saudi Arabia named Saad al-Faqih, who came to the UK in 1994.

When, in December 2004, the United States placed al-Faqih on its list of those having provided "financial and material support to al Qaeda," it alleged that he paid for a satellite phone bin Laden used in carrying out the 1998 African embassy attacks. It said that since the mid-1990s, al-Faqih had "maintained associations" with members of the al Qaeda network, including bin Laden and a key ideologist of violent jihad, Abu Musab al-Suri.

Al-Suri—who was picked up in 2005 in a sting operation in Quetta, the Taliban refuge in western Pakistan—in many ways helped form the foundations of the era's ideological activism. In the mid-'90s, it was al-Suri, then living in London, who first mapped concepts and tactics for undermining despotic Arab governments and attacking their Western sponsors. Many of bin Laden's ideas were shaped by the forceful, intelligent al-Suri, who shaped broad ideological inclinations within the wealthy Saudi that would eventually be honed into a strategy by Zawahiri.

After escaping from Afghanistan to Iran in early 2002, al-Suri—always a bit condescending toward bin Laden, having known the terrorist leader when he was green and confused—began writing a strategic treatise that he hoped would survive him and the other al Qaeda leaders: the "Call for Worldwide Islamic Resistance." The treatise, a sixteen-hundred-page manifesto published on the Internet in December 2004, identifies bringing "about the largest number of human and material casualties possible for America and its allies" as one of the movement's major goals on the path to creating a necklace of regimes across the Middle East,

Persian Gulf, and South Asia linked by the principles of Salafism, the most puritanical strain of Islam.

Al-Suri's manifesto—downloaded relentlessly—has since become something of a planning guide for Islamic radicalism, carrying key concepts from the time preceding bin Laden to some unseen future, when individuals or small groups may have to spread "a leaderless resistance," that will nettle and exhaust the many enemies of Islam in preparation for a head-on battle for the territories that fall under the flag of Sharia. While al-Suri's ideas, which have been widely adopted, map the movement's long-term twilight strategy, bin Laden, since 1998, has been focused ardently on the United States.

"Much of that was Zawahiri's doing," Saad says as we settle on the couches in his study, as was "the very formation" of the bin Laden the world has come to know.

He speaks with a familiarity that comes from having known many of the key players—and knowing them at times of transition, as their defining ideas took hold. It's a view uninfected by personas. Zawahiri, he explains, told bin Laden in 1997 that "Muslims are lacking a leader. They are all waiting for a person who is sincere, austere, living a proper life of piety. And there's nobody who has those criteria. People know that you have abandoned your wealth and joined the mujahideen in Afghanistan, and insisted on staying with them. That you prefer to live in the mountains rather than declare repentance and go live a rich life in Jeddah. And you have charisma—that's important."

What Zawahiri saw, of course, was not just the broader strategy of Salafist jihad—much of it laid down by al-Suri and fellow ideologists—but how neatly bin Laden fit with the ancient narrative of the "wayward prince." That was the key combination, the fusion, where actor and idea became one.

Saad describes all this with soft-voiced detachment, as he fills our cups from a tarnished silver tea service, like someone describing distant history rather than events of only a decade ago that soon after led to tragedy.

He's not strident like the noisy street-corner clerics. Saad is a different kind of werewolf: subtle and scholarly, but every bit as enamored of being in the discussion. He runs a website that has been a favorite of violent jihadists—including the 7/7 bombers—and webcasts a three-hour program a few nights each week into Saudi Arabia. The Saudis, who've pushed the Brits unsuccessfully to hand Saad over, have tried to shut the webcast down. They can't, and Saad—in similarly soft tones—laces into the Saudi regime for its corruption, brutality, and religious illegitimacy. He maintains that Islam endorses women's rights, freedom of expression, and separation of powers—positions that confuse central assumptions about radical Islam's goals. Maybe that's by design—maybe he's carrying forward some elaborate disinformation campaign, sanctioned by bin Laden.

Not likely. His story—of how U.S. responses fit snugly with al Qaeda's expectations—holds together too firmly, too logically, to be fabricated. Zawahiri "understood precisely the cowboy passions of the American establishment" and looked, year after year, to provoke America's native responses. But some of it, Saad says, was dynamic and unpredictable—a process of swift adaptability and, in some cases, dumb luck. The idea "in the first stage" in 1998, he says, was to use the embassy bombings to draw recruits to the safe harbor of Afghanistan. There was internal debate about the targets "we selected," and debate afterward—"when it was clear that many Muslims had been killed and not clear whether that was justified. But then Bill Clinton came on the platform and said: 'Bin Laden's my enemy. Bin Laden, blah, blah. Muslims take him, take your hero.' And then the American politicians and media started cranking up, saying bin Laden is the counterpoint" to America, and al Qaeda recruitment went through the roof. Saad says eleven thousand recruits filled the camps in Afghanistan in a few short years.

The goal behind 9/11, he continues, was "always to create deep polarization between America and the Muslim world." Zawahiri's thinking was that this could be achieved only if America attacked a target in the Muslim world with all its might—as Saad puts it, "in a massive, comprehensive response." To ensure such a response, Zawahiri explained to bin Laden, "you have to attack the heart of America, you have to attack the principles of American might." In this case, those symbols would be economic, military, and political, with the last "being Congress," Saad says. "The target was never the White House." Of course, he adds, America would then have "to answer in a cosmic manner, in a massive global manner, which will force Muslims to remember they are Muslims and do whatever they have to do to stay the jihadi course."

This strategy of drawing responses, of orchestrating them, also took into account, Saad says, a feature of the Muslim personality. He describes this with a story, of a fallen "Muslim in the most sordid place you can imagine. . . . He's drunk; he's seeing belly dancers; he's corrupt. And then he gets into an argument with a cop who swears at the Prophet. The Muslim would then break a bottle and kill him, to restore his identity. 'There, see, I am a Muslim! I am a Muslim, then!'" Saad pauses and sips his tea. "Now, what happens with this individual has happened on a massive global scale."

Of course, not everything went as planned. The swift fall of the Taliban and the elimination of nearly 80 percent of al Qaeda's manpower in Afghanistan surprised both bin Laden and Zawahiri, who expected America to fall into a quagmire as the Russians had in the 1980s. By the middle of 2002, they were both dispirited, on the run, living in caves, with their top lieutenants scattered. "Which is why Iraq was the greatest gift," Saad says. "It proved to the world that it was, in fact, always America's mission to get Muslims, especially when your stated reasons for that invasion were shown to be hollow."

As Saad unfolds the al Qaeda playbook, it's dispiriting to see how well they knew their enemy—the United States—and how obliquely, at best, we understood ours, from Bill Clinton onward.

And the future? Oh yes, Saad says, that's already been written.

Zawahiri calls it "the pacification stage."

Pacification? He nods, this small, bearded man in white. He says, with a gentle smile, that it will happen with something "bigger than 9/11," a WMD attack, that will turn America inward, on itself, such that it withdraws from the world stage. That, after all, was the goal from the start. To both unite Muslims and "collapse the world order," he says cheerily, by making the world's greatest power—"the guardian of that order"—preoccupied with self-protection and internal strife. The analysis is complex, based on the premises that "the lobbies are so powerful in America that they create a kind of intellectual and political authorization—just like McCarthyism" and that "even if you are a wise man, even if you are a leader or a writer, you cannot speak out" and "have to modify your language in order to be in line with the so-called war against terrorism." After the next attack, al Qaeda hopes there will be an uprising, of sorts, of the American people—angry about the country's misguided policies and continuing vulnerability—against the entrenched power of the lobbies and special interests. Summing up the grievance, Saad impersonates the outraged American, " 'What the hell's going on? Why are we forbidding our civil rights, and then nothing happens? We're having more terror, and we're losing money. There's more poverty, more unemployment, and we're losing civil rights. Everybody's arrested for reasons of mere suspicion' . . . and so forth."

It sounds like an overreach, a bit like the sort of talk-radio caller who gets cut off, and I tell Saad so. He shrugs. Right or wrong, it's what bin Laden and Zawahiri see as the goal, as the anticipated reaction, like the string that preceded it, to the next attack. That's enough for them to move forward.

I spend a few minutes challenging these last arguments, as though I'm debating with the men whom Saad seems to speak for, bin Laden and Zawahiri. But it's an argument about the rationale of actions yet to come—actions yet to come in what almost certainly will be a very long battle.

What's clear, however, is that al Qaeda, from the late '90s, was hoping to spur angry, force-based responses from the United States. On balance, they got what they wanted. As they sit in whatever safe house each one is occupying this week in the tribal regions, I wonder what bin Laden and Zawahiri are hoping the United States *won't do*. They are, after all, the lead actors: Zawahiri and his oddly modern creation, bin Laden, hiding among the masses in ragged, dusty clothes. A fabricated but quite convincing challenge to modernity. Both of them—bin Laden the millionaire and Zawahiri the doctor—have actually managed the journey from peak to valley. What they both know—clear in Saad's every word—is that power, both creative and destructive, is locked in the world's restless lowlands, where both men now reside.

What would be the thorn in their side, the thing that would most dramatically undercut the polarization they've sought—that has made the world's 1.3 billion Muslims feel, as most, sadly, do, that they are under siege from America and its allies? Saad acknowledges that the struggle, from the start, is over how each Muslim—at a time of strife inside Islam—defines friend, foe, and the meaningful life. Bin Laden and Zawahiri's plots to spur action and reaction are actually designed to frame that decision—that individual decision—made with relentless repetition. That, Saad says, is what you call "your hearts-and-minds struggle."

But even as he says it, the term seems wrong—wrong in its impersonality—as though vast migrations of hearts and minds move in unison. They don't. It happens one after another—one heart, one mind, at a time.

Dearlove pines for ways to restore our moral authority, Omand for ways to display "our truest values." And what would one person, somewhere, think if America were truthful about what it has learned in these trying years, about its mistakes and its hopes? Would anything change?

Chapter 4
The Trouble with Genies

C andace Gorman stands in the interrogation room in Guantánamo Bay on the morning of July 12 with a document in her hand.

Ghizzawi is sick—as he was in February at their last visit—and weak, and she has already apprised him of the latest set of petitions she'll be filing on his behalf. Since the previous fall—when the Military Commissions Act forbade U.S. courts from hearing habeas petitions regarding detainees at Guantánamo Bay—none of her filings has had any effect, and she expects no better from this set. She's lost her faith in the law providing any remedy for her client. She's an activist now, working outside the system.

But the document in her hand is different from the stack she carries in her briefcase. It's like a living thing. She can't stop reading it, holding it. It's from another detainee case, a certiorari petition to the Supreme Court from late June. The Court turned down an almost identical petition in April—same case, same issues concerning the replacement of habeas rights by Guantánamo's military tribunal system. The difference, this time, is an affidavit—a declaration, really—that the petitioning lawyers collected from an actual tribunal judge. The Supreme Court was so impressed that it reversed its April decision and agreed to hear the case. The Court hasn't done this kind of quick turnaround in nearly sixty years.

Candace tries to explain this to Ghizzawi, how the tribunal judges' names and virtually everything they've done have been kept secret, and how encouraging it is that one of them has decided to go public with his story and honest impressions. Her client is in a weakened state. He sits in his jumpsuit and tries to follow what she's telling him. Wait, she says, she'll read him a section of the affidavit. She flips to the back pages, where the tribunal judge dismantles, step by step, the idea that the CSRTs—the Combatant Status Review Tribunals—abide by any reasonable standard of evidence. The judge cites an example from a case he presided over. His panel concluded that due to the "paucity and weakness" of the evidence provided, "there was no factual basis for concluding that the individual should be classified as an enemy combatant." The panel's decision, the judge adds, was met with resistance from the senior officers who oversee the tribunals.

The filing says that this judge—a lieutenant colonel in the army reserve named Stephen Abraham—served from September 2004 until March 2005, the period in which Ghizzawi's CSRT took place, along with hundreds of others, of course.

"This is a little crazy," Candace tells Ghizzawi, waving the document, "but there's an outside chance this is about you."

On a Monday morning a month earlier, Stephen Abraham was sitting in his office in Newport Beach, California, going about his business. He's a neatnik—no piles on the desk—with everything for the coming week neatly stacked on a side table: a filing against a lending company he represents, another one against a landlord with some angry tenants, and an action against a pomegranate juice company, a little local one, being sued by the giant "Pom" juice conglomerate, Fijiwater, which Abraham contended seemed to claim to own the very concept of squeezing pomegranates. In his court filing, one he was particularly proud of, Abraham cited Bathsheba from the Bible—"I would cause thee to drink of spiced wine of the juice of my pomegranate"—wondering if Fijiwater claimed ownership over her as well.

Abraham is a professional nuisance, and a good one. So is his sister. She's a lawyer, too, at a big firm on the East Coast. They're children of a Holocaust survivor, an immigrant to the United States who spent time in labor camps as a boy, and both of them are a bit zealous about parsing truth from deception, and then not budging.

His sister was, at this moment, calling him. He could see her number on the caller ID, and he knew what she wanted. But he wasn't sure he wanted to pick up. It was about his old life, one he was trying to leave behind so he could get on with just living.

But he did pick up. "Okay, okay, but how long do you think this presentation is going to go?" he asked.

"As long as you want to listen in," his sister, Susan, said. "I thought it'd be interesting for you."

He signed on to his computer and clicked on the link she had sent him, which opened a videoconference being broadcast for all fifteen offices of Susan's law firm, Pillsbury LLP. A lawyer from the firm talked for a few minutes about the firm's pro bono work representing detainees, how Pillsbury got involved and how the process of adjudication worked at Guantánamo.

Abraham watched and felt, minute by minute, that he was being drawn backward into the churning surf.

Abraham left active service in the army reserve in March 2005, after completing his term with the Office for the Administrative Review of the Detention of Enemy Combatants, or OARDEC. Actually, he resigned to keep from being fired.

It's hard to imagine things turning out much differently. Abraham was misplaced at OARDEC by virtue of both his experience and his inclination. He knew too much walking in the door. He'd served as an intelligence officer in active and reserve capacities since 1982, including a full-time deployment during Operation Desert Storm. After 9/11, he was assigned to Pearl Harbor, where he worked as the lead terrorism analyst for the joint intelligence center, Pacific Command. Then, a couple of years later, the call came from OARDEC, and off he went to Washington.

It was soon clear that he was one of the only intelligence professionals in a building filled with lawyers, all of whom had to answer to officers being watched warily by the Pentagon. Paul Wolfowitz, the deputy defense secretary, and Undersecretary of the Navy Gordon England seemed to have placed Geiger counters inside OARDEC. And with good reason. Both men, and their superiors, knew that Guantánamo sat on a fissure between American ideals and actions, and that year by year it had become a symbol of how those geological plates were separating.

By the summer of 2004, the pressures under Guantánamo had become acute. The Supreme Court's late June ruling in *Rasul v. Bush* essentially restored habeas rights to the detainees and opened the door for them to challenge their imprisonment in U.S. courts. The administration scrambled to put in place a surrogate review process that could fill in for habeas corpus while keeping everything in the hands of the military—a process that would look like justice but would keep all goings-on in the world of the "classified." Thus the CSRT program was born. Barely a week after the court's ruling in *Rasul*, Wolfowitz passed the order to Gordon England, who publicly announced that all 558 detainees would be put through "status reviews" within six months, a wildly unrealistic goal for the completion of any evidence-driven proceeding.

Into this realm stepped Stephen Abraham. Due to his intelligence background and expertise, he was placed in a pressure-point job: managing the flow of available evidence, or "information" as he'd call it, to the CSRT hearings. This meant he got to see how much, or how little, the decisions of the tribunals actually rested on.

Things came to a head when, as a reward for his instrumental work, he was asked to sit on one of the three-judge panels. With him were a man and a woman, both in the air force—a colonel and a reserve major in the JAG Corps, respectively.

This was November 2004, the twenty-fourth. The detainee was a Libyan national picked up in Afghanistan in late 2001, interment serial number 654. His name was Ghizzawi.

The procedures that framed the CSRTs had a strange disconnected quality, like a tax audit on a fictional character. A "recorder"—often a young officer— would present the panel's judges with a narrative of the case. The recorders were a bit like scribes on the rewrite desk at a newspaper—reporters would call in the facts, and the scribes would weave them into a story. Also present at the tribunals was a "personal representative," a sort of liaison to the detainee whose case was being reviewed. This representative typically had little contact with the detainee and was more an observer than an advocate, which left the process without any of the adversarial pressure that drives traditional legal proceedings. But Abraham would provide that, just as judges often do in countries with non-adversarial systems, such as France.

Pity the recorder. Abraham understood his job better than the recorder himself. The evidence arrayed against Ghizzawi, as with many of the detainees, was a lot like the images in a hall of mirrors, warped and skewed at various angles but all depicting the same single object. The object, in Ghizzawi's case, was his admission that, while in Pakistan in 1997, he attended a meeting of the Libyan Islamic Fighting Group and signed up as a member. In those days, the group, which opposed Moammar Gadhafi, was seen favorably by the United States and served as a magnet for expatriated Libyans—much like an IRA rally in the 1970s might have attracted someone from South Boston. This was something that Ghizzawi had "offered voluntarily," the recorder stated. Abraham had regularly seen this language, especially when detainees, often under torture, offered innocuous information—many times no more than a single fact—that was then multiplied into a condemning tableau of speculation and suggestion. In the case of Ghizzawi, the recorder placed him at virtually every event and location in Afghanistan that could be linked to the Libyan group, and the Libyan community more generally, including a guest house where Osama bin Laden had stayed for a night in 2001. Ghizzawi, potentially anywhere and everywhere in Afghanistan that angry Libyans were to be found, across several years, was alleged to have provided security for bin Laden that night. The designation "acted as security guard for Osama bin Laden" is a charge so often leveled at detainees that it has generated a joke among Guantána-mo's military attorneys, as Abraham later recounted: "Bin Laden should be easy to capture: just look for a roving band of five hundred bodyguards from twenty-two countries, and you'll find him, dead center."

As Abraham picked apart the recorder's narrative, it soon became clear that the government's case against Ghizzawi relied on penumbras emanating from a tiny square of paper. The scrap had been found in Ghiz-zawi's pocket, the government alleged, and bore the phone number of another detainee—though not, inci-dentally, a detainee about whom there was significant evidence of wrongdoing. When questioned about the piece of paper, the recorder stated with dark imputa-tion, Ghizzawi could offer "no explanation." Abraham followed up, asking if anyone had actually asked the detainee for an explanation. The recorder balked. He said he wasn't certain if anyone had, but he supposed he could check. While he was at it, Abraham added, he might as well look into the "chain of custody" issues concerning the slip of paper. There was nothing more to present, and little more to say. The panel unani-mously recommended that Ghizzawi be reclassified as a non-enemy combatant and begin the process of release and resettlement.

The decision sent a hot current up the chain of command. Soon, OARDEC's director and deputy director—retired navy officers James McGarrah and Frank Sweigart—were weighing in. There were questions the panel had requested the recorder answer. Wait for those answers, they told Abraham and his fellow judges. A week later, the panel reconvened to receive them. There was, in fact, no record of Ghizzawi having been questioned about the slip of paper. About the matter of chain of custody, Abraham was particularly attentive. He'd been a Hollywood, California, police officer for a brief period in his twenties. The slip of paper would fall under the official category of "pocket litter"—that is, whatever is in a suspect's pockets at the time of his arrest—which is generally seen as low-priority junk and is often mishandled. He pressed the recorder on this point, until the recorder acknowledged that there may have been some confusion in the handling of the pocket litter gathered from groups of captives in Afghanistan. But even if there was irrefutable proof that the piece of paper with the phone number had been in Ghizzawi's pocket, it would prove nothing, Abraham felt. The other detainee mentioned on the paper, Abraham said, could have been "a guy he'd met a few days before, on the street, or someone standing next to him in a cell."

The panel reaffirmed its original finding: non-enemy combatant.

At this point, Abraham was tagged as trouble; with his rigorous approach and evidentiary standards, he could bring down the entire structure. This seemed only to egg him on. Taking to his absurd task of validating the nonexistence of exculpatory evidence in the detainee cases, Abraham made trips to Langley, to query CIA officers about the origin of various pieces of intelligence. This was familiar turf for Abraham, and he knew too much to be fooled. It had never been in his nature to keep quiet. The meetings were frustrating. The intelligence officers were not forthcoming about the source of their information, and as Abraham pressed, it became clear, hour after hour, how little real evidence there often was.

On December 10, he typed up a resignation letter and passed it to Frank Sweigart. There were several exit meetings in the coming three months—where his ongoing obligations to secrecy were stressed—before he returned to California and his law practice, Fink and Abraham, with its two offices and no secretary, overlooking a parking lot.

Which is where, two years later, he sat on June 11, 2007, watching the Pillsbury videoconference and trying not to think about the pile of work that beckoned: a loan company to defend; pomegranate wars.

Abraham once again knew too much. The Pillsbury lawyer seemed to be reading from a manual about OARDEC's procedures, about the military's protocol for gathering evidence and the status reviews for the detainees. He grabbed his BlackBerry and tapped in a note to his sister: *An A for effort, but your guy has no idea what he's talking about.*

It went on from there—a BlackBerry fusillade, note after note. *Ridiculous. Absurd. Let me give him directions to Guantanamo.*

He had nothing against Matt MacLean, a good lawyer and former JAG officer doing his best. It was more like watching a tourist lose his shirt at the blackjack table when you knew the game was fixed.

Afterward, a call came from David Cynamon, Matt MacLean's older partner, and the firm's lead lawyer on a key Guantánamo case, *Al Odah v. United States.* The case, filed on behalf of a group of Kuwaiti prisoners, had been mired in the same stalled habeas petitioning as all the others since the Military Commissions Act. In April, the Supreme Court had refused to grant *Al Odah* certiorari.

A few days before the videoconference, when Susan

was signing up to participate, she essentially outed her brother, telling Cynamon and MacLean simply that Stephen had been a Guantánamo judge and hadn't been pleased with his experiences.

Cynamon now moved gingerly on the phone, mentioning that when the petition for certiorari was turned down in the spring, the Supreme Court had relied on a declaration from former OARDEC director Admiral James McGarrah. Abraham said he knew nothing about any of that, that he'd left detainee issues far behind him, didn't even the follow the pertinent stories in the press. Undeterred, Cynamon simply wondered if Abraham had ever read what McGarrah had said in his declaration about the workings of Guantánamo. Maybe Stephen could glance at it and pass along any insights. Abraham paused. "I'm not sure if I'll have time to read it, but you can send it along."

The .pdf file didn't sit in his e-mail in-box long. The downloaded document staring at him was McGarrah's official version—the agreed-upon story of Guantánamo—and Abraham couldn't resist a quick look.

When McGarrah explained that the CSRT could "request the production of such reasonably available information in the possession of the U.S. Government" concerning a detainee's status, Abraham punched in a first comment, an annotation that was lawyerly in nature: *Begs the question of what was reasonably available.*

It didn't take long, however, for Abraham to engage more forcefully. Each recorder, McGarrah wrote, was "charged with 'obtain[ing] and examin[ing]' the Government information."

Recorders had no clue as to what they were holding, Abraham shot back. McGarrah went on: "The dedicated Team focused on the tasks of identifying relevant information on each detainee." Relevant?

Without understanding the complexity of the issue from a legal or factual standpoint and with no background in the subject area, the "team" could not possibly know what was relevant.

"Members assigned to the Team each received approximately two weeks of training prior to assuming their data collection responsibilities," McGarrah explained, including "intelligence training."

Annotation: *After 25 years, I don't begin to profess any expertise on the subject. After 2 weeks, they're experts?*

"Quality Assurance" reviewed the case materials, McGarrah stated, "to ensure they were logical, consistent and grammatically correct."

With emphasis on grammar! Abraham countered.

The recorder, McGarrah continued, "could add information . . . that might suggest that the detainee should not be designated as an enemy combatant." This was a long-standing sticking point with Abraham— proving the negative. How could this be possible when detainees so often fell far below the threshold of even modest relevance? *According to the Admiral, our systems include data on the 5 billion people that aren't terrorists?? Negative intelligence works for weather, not human assessments.*

The contrapuntal dance of McGarrah's sober assertions and Abraham's incredulous rejoinders soon littered the document. The admiral's testimony relied on words of reassurance: *reasonably, relevant, quality.* The statement wasn't so much false as hollow. "The Team and Recorder," McGarrah wrote, "ensured that, as they reviewed Government Information, *all* material that might suggest the detainee should not be designated as an enemy combatant was identified and included in the materials presented to the CSRT and included in the CSRT Record."

Abraham read the summation and sat for moment, trying to bottle his combative urges. He was a lawyer, and an intelligence officer. He knew what he knew from personal experience, hard earned. This wasn't about McGarrah or about Abraham's anger about how he himself had been treated. This was about OARDEC, and what happened there. He punched up the annotation screen.

The limitations relating to information collection assured that the body of information was never more than grossly inadequate.

The story of OARDEC in fewer than twenty words.

He sat back in his chair and took the phone off the hook. He knew that if he hit the Send button and shared the marked-up document with Cynamon, everything would change. He'd be handing Cynamon a weapon—a tribunal judge challenging the former head of the entire tribunal system—and he'd be drawn back into the ulcerous fray. He didn't think he was up to it, not anymore.

After 9/11, he had become single-minded and obsessive, a middle-aged warrior. His marriage of ten years ended. His daughter went to live with her mother as Abraham vanished into the new and consuming counterterrorism fight. He'd always been fiercely patriotic—a know-it-all, unrelenting and in-your-face. When his father came to America, he had had nothing but the rags he was wearing. The country had provided refuge, and his son, especially after the Trade Center towers fell, wanted to return the favor: a Jewish avenger, for as long as the battle raged. But then things turned sour. Abraham's notions of justice and fairness seemed secondary to some tactical plan from on high, one that eluded him. He'd never been as angry and dispirited as when he left OARDEC. The job had devoured him. By the time he walked out, there was little left to be consumed.

But over two years, he'd crawled back. He had a good business, a new fiancée. "No, thanks," he thought, moving the cursor off Send. It was time to look toward the future. What's past is past.

He turned his attention to civil litigation and pomegranates for a few hours, as if nothing had happened. But then he found himself looking at the key, a large ornamental key, resting on the corner of his desk. He'd bought it on Christmas Day 2004, when he took his ten-year-old daughter, Rachel—visiting him in D.C. on her winter vacation—to Mount Vernon. He and Rachel were walking through the empty mansion when he saw it, affixed to the wall in the main entryway: the iron key to the Bastille, about eight inches long, a gift from the Marquis de Lafayette to Washington. Abraham, at that point, was a mess—just two weeks before, he'd handed in his resignation letter. But his daughter asked about the key and he summoned his usual jaunty cadence, telling her about the Bastille, about how the Parisians stormed its gates to start the French Revolution and found only a few toothless beggars inside. Then father and daughter huddled in close and read Lafayette's inscription on the adjoining plaque: "Give me leave, my dear General, to present you . . . the main key of the fortress of despotism. It is a tribute which I owe as a son to my adoptive father, as an Aide-de-Camp to my General, as a Missionary of liberty to its Patriarch." Rachel asked her father what all that meant, and Abraham told her it was "about democracy's promise to always free the innocent," and then he thought of his own father. And he turned away so Rachel wouldn't see that he was overcome. On the way out, he bought a replica of the key at the Mount Vernon gift shop—four inches long, gold-plated, in a little glass case—for $10.99. That key, which now sat on the edge of Abraham's immaculate desk, reminded him of all the most important things, tucked, right there, in the mix of what he'd been trying to forget.

He hit Send.

Abraham's many annotations on McGarrah's declaration, and an hour-long conversation with the Pillsbury lawyers later that day, would form the core of his own declaration, filed to the Supreme Court before the week was done. It said, in essence, that the Guantánamo system was built, maybe intentionally, to fail. Either way, the most basic standards of justice were not being met. And he was a credible witness.

Two weeks later, the Supreme Court granted certiorari to a merged case—Cynamon's *Al Odah* folded within a similar case, *Boumediene v. Bush*—that would present a full-frontal challenge to the suspension of habeas corpus. At the end of July, Stephen Abraham would testify before Congress: know-it-all heaven.

The case, due to be argued in the Supreme Court's following term, would allow a kind of constitutional showdown between the branches of government over fundamental principles. Such moments create the opportunity—available, if not always seized—for America, both government and people, to integrate its many parts, its many urges, around shared and sacred principles.

The exercise of power—crucial to progress in many ways—is actually not one of those principles, though it's adept at pretending otherwise. Across millennia, great empires have amassed and projected power, sometimes brilliantly, giving rise to centuries of primacy and lasting influence. And who wouldn't want that?

But that's not what makes America unique in the long human pageant. It is, in fact, a wildly improbable inversion of the way the world has long been. The official record, in the summer of 2007, shows that a man with no power, naked and freezing on the floor of a cell, managed to marshal the concept of presumed innocence and the conscience of a Holocaust survivor's son to crack open authority's gilded vault.

Because Abdul Hamid Abdul Salam Al-Ghizzawi's lone advocate, Candace Gorman, is right when she waves the document in front of his face on a hot July afternoon.

It *is* about him. All of it.

The Wild Hogs rule!

Ten bikers strong, moving as one, the big Harleys scream down Route 1, the coast road, on a high cliff above the heaving Pacific surf.

At the center of the snaking chrome: Jordan's King Abdullah and Rob Richer. Only one actually *rules.*

They saw the movie *Wild Hogs* last month. HM— Rob's shorthand for "His Majesty"—loved it, laughed like hell. That Tim Allen guy. And don't get him started on Martin Lawrence.

They'd decided, why not? "You're the king," Rob said. "Let's do it."

The morning mist, salty with ocean sprays, mixes with Harley fumes as they race south, two hours from San Francisco, on their way to a roadhouse in . . . *Carmel?*

The gang—with four security vehicles, two in front, two in back—rumbles by boutiques on Mission Street in Carmel-by-the-Sea and into the parking lot of Katy's Place, a cute little breakfast joint with a fenced-in patio and impatiens—red, pink, and white—blooming in the window boxes.

Alerts were quietly sent up and down the coast, everything kept very low key. A Secret Service detail in shorts and polo shirts has already checked out the nail salon, the art gallery, the Western apparel store. Three dark-suited agents then take up their posts at the restaurant's entrance, as Jordanian Royal Guards, working with the California Highway Patrol, clear out the place—no great feat at 10:45 on a Wednesday.

The entourage settles into their tables. HM and Rob—both in fresh leather, the shiny bikes leaning outside in the sun—sit together with ten others at the big table. The king goes with the Swedish pancakes. Rob—a John Travolta to Abdullah's Tim Allen—goes with the corned beef hash, lots of Tabasco, a biker order that definitely *rules*.

They're making each other laugh and feeling very good. The king "needs his space," a Jordanian official traveling with the group says; the trip is "extremely private." And that's understandable for reasons beyond personal preference: the king, a child of Hussein's English wife, his first wife, has been looked upon with a circumspect squint by some Jordanians. He went to exclusive Deerfield Academy, in Massachusetts, attended Sandhurst, in the UK, for military training, and has worldly tastes for a member of the Hashemite family, one of Islam's founding clans, and a direct descendant, forty-three generations along, of Muhammad. In a country that's 90 percent Muslim, with a growing fundamentalist movement, he has to exercise some care with regard to appearances. Harleys and leather are not preferred accessories.

At least, not in Amman. But in California, a land built on self-invention and the future, he's a natural addition. Of course. A Middle Eastern king is here on a Harley—looking, with his trimmed beard, like a movie villain or, well, a Middle Eastern King. What else would he do? The co-owner of Katy's, Randy Bernett, hands out boxed meals to the Secret Service guys guarding the door—"they got gypped out of the best breakfast in California"—and strikes it up with Abdullah. Bernett's a faithful reader of *The Economist* and has just bought a new DVD version of *Lawrence of Arabia*, which includes in the bonus features footage of King Hussein watching the movie's filming from a director's chair. "I really appreciate the job you're doing with the peace process over there," Bernett tells Hussein's son. "Getting involved the way you are could make all the difference."

"Thank you," Abdullah says in his soft, formal voice with its vaguely British precision. "I don't actually hear that very often. I quite appreciate it."

The other proprietor, Randy's ex-wife, Gytha—a blond beauty from Iceland—also talks a bit of geopolitics before giving HM an affectionate hug. Word has by now spread that royalty is in town, and one hundred Carmelites are outside applauding as the Hogs roar off, bound for Big Sur and four days of adventure. Arnold Schwarzenegger quietly slips out of the governor's mansion to dine with the riders, as do several other California notables, and a few days later, they meet up with the former Miss Jordan.

Rob is always at HM's side. That's right where he stays when both men return east over the weekend. Abdullah meets with Condoleezza Rice on Tuesday, and then dines with Bush and Laura that night in the White House residence.

The conversation is predictable. Abdullah is trying to lure America back to its traditional, honest broker status in the Middle East region. He says that America's focus on the use of military power—now mirrored in Israel—is demonstrably misguided. The fabric of the region has been badly torn and real diplomatic outreach—with the United States acting as a stabilizing rather than disruptive force—is the only hope to mend it.

And an opportunity to change course is at hand. Last week, Bush called for an international sit-down of Israelis, Palestinians, and their Arab neighbors to work through issues that have long stymied peace negotiations and the establishment of a Palestinian state alongside Israel. The UN General Assembly is meeting in New York in September; they all can talk there.

Abdullah, in fact, spoke with Bush from Southern California and cut the Wild Hogs adventure a few days short to meet with the president and discuss negotiating strategies. Over dinner, he tries to stress what will work, how Bush needs to show some sympathy with the Palestinian plight, maybe to acknowledge that some of the moves up to now, by the United States and Israel, at the very least, didn't result in the intended outcomes. He's trying to encourage Bush to admit error, which the man is almost chromosomally incapable of doing. He sees it as weakness. Bush believes in will, in resolve, and in their power.

But as interesting as anything said that night, Abdullah finds out that Rumsfeld and Douglas Feith were in the White House over the weekend, advising Bush. "Are they still advisers?" Abdullah asks Richer, darkly, as he leaves for Jordan. "I thought he was done with them."

The next day, Richer is ordering lunch at the Ritz-Carlton in Northern Virginia. Like much of the rest of official Washington, he's dispirited about the U.S. government's lack of direction and its simplistic, ineffectual responses to complex problems.

Rob talks for a bit about the Jordanian king's visits with the president and the secretary of state—and then takes a phone call from his son, a teenager, who's off with Abdullah's son at "an adventure camp."

"The U.S. government is broken, unable to budge, or save itself," he says when he gets off the phone. "That's why everyone is moving unilaterally, on their own. Abdullah is trying to be a one-man 'honest broker' department, to be the referee, in a region where the big honest broker, America, is on the sidelines coaching Israel. Someone's got to do it. . . . 'Cause, really, what's happened . . . some people, like Abdullah and a few others, have stepped in to fill the vacuum we've left 'cause we're not engaging the world."

Richer then jumps into his personal mission, his way of engaging with the world. Of course, it's operational. The can-do guy's campaign to "shock the beast."

He's been working on it since the spring, step by step, on picking teams.

"I've got the infrastructure set up, and I just used it to do something else, which I can't talk about. I'm dealing with Mafioso in Sicily, diamond smugglers in Antwerp. My principals are, in one case, a former CIA officer; in another case, a former French Legionnaire who then worked with French intelligence. I know the three principal guys who report to me and know me by name. Only one of them would I ever be able to put somewhere and say, 'There is this noble goal to this operation.' Because they then go, and the guys who would get the stuff—the uranium—actually, they don't care who they're selling it to. So here's what we gotta do . . ."

And he's off, in operational mode, sketching it out, with twists and turns, trapdoors, and high-risk maneuvers, until he delivers "one hundred pounds of HEU to the steps of the Capitol. Then everyone will know what's real."

Experts on Rolf's IND Steering Group are convinced that a likely circumstance is that a terrorist network will smuggle the enriched uranium—which gives off a very low level of radioactivity—into America and set up the workshop for milling it into a bomb-ready form and final assembly of the device inside the United States.

In the spring, Richer worked out those logistics. "You need to know the right airports to go through. You want to come into the South . . ."

He then describes how he had an operative—a double amputee with two prosthetic legs—board a plane in Austria, bound for Atlanta, Georgia. "He's got an army T-shirt on. And he's wheeled in by the guys up to the gates in Austria. They put the wheelchair through the thing and have him walk around it. They don't scan him 'cause everything's metal.

"He arrives in the States. He comes off the plane. Once you're in the States, you don't go through any security. So he never got X-rayed. And even if they had, they wouldn't see a thing. They can't get into his prosthetics. They can't find the seals on it. *They can't find my seals.* If they were to say, 'There's something inside there?' My guy would say, 'Yeah, shock absorbers.'"

But they didn't do that once he got to Atlanta. He sailed right through. In one prosthetic leg was a significant amount of lead powder—which represents the HEU, which is often in powder form—and in the other was scrolled paper. "I wanted to see how I get documents in. Actually, it was blank paper. If someone managed to take his leg apart. He could say, 'Some idiot, when he made these things, put paper in it.'"

Rob stops, like he's snapping out of a trance.

"This grouper looks great!"

It is great, caught in Florida, quick-frozen and shipped to a Northern Virginia restaurant, in a vast, big-hearted nation that is impossible to defend. Now ordering tiramisù for dessert, Richer has arrived where the most knowledgeable people in Homeland Security end up: recognizing, often against their will, that the forceful protection of America's endless borders is untenable. In the man-on-man war, defense doesn't work. Too many men destined for too many attractive targets.

Across town, Rob's longtime partner, Rolf Mowatt-Larssen, is still working on two tracks: inside and outside. Outside is now just he and Rob, and their scheme, their "shock to the system" test.

Inside? A solution from the inside—finding a way for the U.S. government to see within a secret world of malevolent buyers and amoral sellers of uranium—is his last best hope for public service. That's the way he thinks about it.

He's giving it his all. He's been taking the IND Steering Group—Vahid Majidi and the rest—on trips all over the planet. In the early summer they ran a loop through Britain, France, and Russia. This falls under the category of "joint"—joint multilateral initiatives where intelligences services work together on this one issue: terrorists and nukes. It is, of course, too much to ask that they work with unity and shared purpose on other issues. They're competing teams. They spy on one another. America spies on all of them. All of them spy on America.

But Rolf, in each foreign capital, said, "This one area is different, and we need to figure out a way to break down the walls and create a separate safe room where we all can gather." In that safe room—and there would be one "joint office" in each country—the Russians, French, British, and Americans, at first, would share all information on terrorists and WMD; they'd create cover companies, run operatives, tap officials in one another's governments. Everything would be shared. "Think of it as the state-based solution," he said. "It's us, the states, against them, the transnational terrorists."

Everyone thought this was a splendid idea, but no one had any clue how to write the rules. Would we run joint undercover operations, the British asked, inside each other's countries? What if we have sources that might be helpful in ferreting out terrorists on the WMD hunt; would those sources, the French queried, be revealed to all parties? And then there were the Russians, who wondered if someone was found smuggling

HEU "inside one of our countries—and it could be the U.S. as easily as us"—would it have to be made public, embarrassing that government? Would that be a joint decision, or would one country have veto power?

Not that the British were anything less than enthusiastic and civil, or the French anything but hospitable and frank—they all dined at a château. The Russians—well, they were more open than usual. Vahid gave a speech, saying, "I'm in a kind of job where all of you should want me to succeed," and the Russians laughed, and the vodka flowed that night.

The joint multilateral effort is just one worthwhile endeavor. There are others inside the government and stretching across the globe. None of them is expected to change much. At best, they'll offer slow progress and modest returns.

To keep sane—to feel like he's doing everything he can—Rolf has pressed CIA several times during the spring to launch its own teams. For them, he calls it the A. Q. Khan approach—a term that carries currency inside the agency. The takedown of A. Q. Khan, still prized at Langley as the era's finest clandestine work, was the result of a patient, relentless eight-year operation. Khan, in the late 1980s, began to vigorously sell the blueprints for uranium-enriching centrifuges and plutonium-reprocessing techniques to the North Koreans, the Iranians, and others. At the time, he lorded over an array of specialty manufacturers who could produce and provide the equipment. That's where CIA began its infiltration, placing agents and developing sources within these manufacturers. The agency closed the net, bit by bit, year by year, until Khan was exposed and put under house arrest in early 2004.

The "teams" could be similar, moving through high-probability areas, looking for the sale of HEU; they'd have covers—as companies or criminal syndicates or polished and capable jihadists—that would become more familiar and accepted as the years passed; over time, they'd build sources and contacts. It was a matter of planting seeds and letting them grow. One ace in the hole was that the man who'd led CIA's A. Q. Khan mission, Jim "Mad Dog" Lawler, now worked for Rolf as a contractor. Rolf and Jim together could guide a joint project with CIA, which had a clandestine capability that Energy still lacked.

Steve Kappes, the agency's deputy director, and other CIA bosses were intrigued, but skeptical. What if the team got caught? That was always a first concern. Does the United States deny ownership, and would that be credible? The embarrassment factor cut both ways: if you happened to be the country inside of which the HEU was bought, or if you were the country, the United

States, whose team was arrested or, worse, killed. The key to the A. Q. Khan mission, they countered, had been A. Q. Khan. The United States had been duly suspicious of his activities since the early 1990s. That mission had a target. What was the target here: some shadow smuggler or jihadist buyer of fissile materials, yet to be found? Kappes and the others admired Rolf's ingenuity—yes, he was still thinking a few steps ahead of the pack. But this, one CIA chief said, "was a fishing expedition, and a high-risk one—that's not us anymore." What he meant was that such an operation was no longer suited to the current means and methods of an intelligence community, tethering, ever more tightly, to a military model: attainable goals, linear execution. Check the box, move on.

By midsummer another challenge was coming from inside the IND Steering Group. The term of art was *market stimulation.* Undercover teams with thick wallets, out looking to buy HEU, might "stimulate" the markets, creating activity where it might not have been before. Vahid Majidi, from FBI, was especially concerned about this. Others, as well. It's a common conundrum for sting operations in general. Do they compel criminal activity that might not otherwise have occurred? After an arrest, defense lawyers will claim entrapment—that a defendant was lured into the activity in question—but courts tend to be unimpressed by that argument. In this case, Majidi felt, potential suppliers might be lured into the market if they thought a real buyer was on the hunt. Rolf countered that they already knew there were real buyers—including bin Laden—out there. Luring potential suppliers into the market—be they some former Russian general or an agent of North Korea's government—was, in fact, a goal. Around they went, recognizing that creating teams and launching them might burn off some of the so-called ambiguities—ambiguities, in the mind of Majidi and others, that created a protective confusion about who was a real buyer and who a real seller.

"But that's going to happen anyway at some point," Rolf said, in one IND Steering Group meeting. "Let's manage it so at least we can own that process rather than be surprised by it on some catastrophic day."

Meanwhile, what Rolf and some at CIA suspected—and what several thin intelligence reports were starting to confirm—was that the markets were already evolving swiftly, in terms of both buyers and sellers. Why? A new entrant? Actual buying teams from states that had moved from have-nots to why-nots on the subject of nuclear weapons?

The logic is airtight for some modest-size state in the Persian Gulf—or Africa, or the Balkans—to say, "All we need is one bomb in a warehouse with a delivery system." It would be there, lying in wait, for defensive purposes. With widening membership in the nuclear club, which now includes North Korea and soon possibly Iran, how could a state—any state—afford not to have defensive nuclear capability? A ten-kiloton device like "Little Boy," the Hiroshima bomb, would be plenty. But how do you get the fissile material? Enrichment is expensive and almost impossible to hide, with all its cascading centrifuges. Let the Iranians fight that battle. All I need as a "why-not" country (for example, Saudi Arabia or Sudan) is to have my intelligence service send out an "unmarked team"—exactly like one of Rolf's teams but with the opposite end goal—to buy, or steal, the requisite uranium. The team would have real money and the ability to verify the authenticity of the materials—resources and capabilities that would squeeze ambiguities from the markets and, yes, stimulate activity. The United States knows from hard experience that when anything makes this much sense, someone is trying it, and the government is searching through the summer of 2007 for these unofficial state-based buyers. They have to be out there. No hits yet.

It's the multiplying of these unanswered questions and the general vexations over who should be in charge of this terrain that prompts a wide-ranging turf battle by late July. *Turf battle*: that hard, ugly expression of craven, self-protective bureaucracy. But pull it apart in this case, and there's a soft, familiar core—confusion. The wrangling over turf within a government is, after all, often the result of disunity in overarching principles, a sign of a government not being integrated. The many arms of a government are left in a state of disarray, in a free-for-all, and they fight it out.

The conventional remedy? A meeting with the president. Let the president take ownership, set a direction, maintain order in the ranks.

While Bush goes on his August vacation, Rolf and the directors of other intelligence agencies, or their deputies, or their terrorism chiefs (almost every agency has one now), or their WMD specialists, huddle. The meetings are a slowly unfolding mess, each small step a negotiated compromise. Rolf has been saying for a year, in front of every commission, in countless meetings, that the United States needs to have a czar, someone with direct access to the president on the issue of nuclear terrorism. This advocacy, within the passive-aggressive illogic of bureaucracy, made it such that he—already in that role, ex officio—couldn't nominate himself.

It is decided, nonetheless, that Rolf should give the presentation to Bush. Everyone knows the two men have a history together. Rolf is the natural choice—but he must be controlled. That's the thinking among the heads of the other departments.

The quid pro quo is that, yes, he can be their representative, but he must truly represent a wide array of groups, from FBI to CIA to DHS to DNI. All of his hard-earned beliefs are squeezed through a filter of departmental self-defensiveness. And the fight is on.

We have no intelligence penetration of al Qaeda or intelligence presence in the markets. You can't say that. It reflects badly on CIA, and we do have some projects in the planning stages, some Rolf himself has suggested.

The U.S. borders are porous, vulnerable to anyone looking to smuggle fissile material or a nuclear device into America. Now, wait a minute. DHS, with its host of new radiation detectors, is certainly on the case.

And on it goes, until the last few days of August, when the president returns from Crawford and begins to settle back into his daily rhythm.

Rolf sits with Energy Secretary Sam Bodman, a former chemical engineering professor at MIT who eventually became CEO of Cabot Corporation, a Boston-based seller of specialty chemicals, where he made a fortune. Bodman, a quiet, cerebral man who jumped between the deputy secretary jobs at Commerce and Treasury before moving to Energy in early 2005, has become increasingly zealous about the threat of nuclear terrorism. He and Rolf have, at this point, spent a great deal of time talking through the subject. Though he hasn't worked as a professor since the 1970s, Bodman still appreciates a good freewheeling discussion of theory and practice. In this case, he is particularly attentive to the lack of hard evidence about the disposition of the world's fissile materials. He understands the science and how readily a team of modestly trained physicists and engineers could build a nuclear device of significant yield. He's been wanting Rolf to sit with Bush for a long time. Now he's helped set up a meeting of NSC principals, along with the president and vice president, for August 28, two weeks before the 9/11 anniversary.

"Well, look who it is, Mr. Bad News," Bush says, as he spots Rolf. "Every time I see you, there's something terrible to talk about. But I guess that's good—better I hear."

They shake hands.

"How ya been, Rolf?"

"Fine, Mr. President, and yourself?"

"World still in one piece?" the president asks.

"Yes, sir, I suppose it is."

"I guess we must be doing something right," Bush

says, as the vice president and other NSC principals take their seats. Rolf has slides prepared—he knows Bush likes the PowerPoints—and he runs across the harrowing landscape. The presentation is cleverly constructed. During meetings Rolf attended in 2002, Bush at one point said, "How can we find the Atta?" It became a favorite phrase of Bush's. The slide "Unique Challenges" starts with "How can we find an Atta-like footprint?"

Bush nods. "It's small footprint, isn't that the problem?"

Yes, that is the problem, Rolf says—the problem of the modern age, where ever-growing power can be seized by ever-smaller groups, with tiny footprints—and he runs through a basic six-point action plan: Penetrate terrorist plans and intentions. Get nuclear material off the black market. Ensure the security of all nuclear material worldwide. Manage nuclear material information. Lead global cooperation to track and interdict the smuggling of nuclear material. Establish a coalition to deter nuclear terrorism.

The six points—so much more modest than the mission to rid the world of evil that animated Bush and America after 9/11—now seem thoroughly, sadly, out of reach. The cats are not being herded. They're running free, everywhere, as a diminished America struggles to regain its footing. It's something everyone in the room knows, but no one can talk about, and the slides flash forward, all the way to a final slide, "Are We Up to the Challenge?" and three bullet points: Never Underestimate Your Enemy. Know Your Enemy. Know Yourself.

Bush is agitated, inquisitive, and hyper-engaged like he was in his big years, his years of action—the first two after the attacks—when Rolf always talked straight to him about the terrorists' intentions to get WMD, about their progress in doing so, about the chemical attack on the subways that Zawahiri called off in favor of something bigger, about al Qaeda's hair-raising nuclear and biological programs.

"So what's our plan, what's under way?"

Rolf pauses, gathers himself, and moves into the agreed-upon program, a précis of what each department in his coalition is doing: best-laid plans of CIA, the radiation-detection programs of DHS, FBI's new strength in accounting for the origin of stolen uranium, and the joint intelligence initiatives he's crafting with the Brits, the French, and especially the Russians.

Cheney asks, "How cooperative are the Russians being, really?"

He knows Rolf's a Russian expert, and the exchange is frank: the Russians really don't buy the threat of terrorists with nukes as a crisis. "The only thing they'd see as a crisis," one of the principals says, "is oil at less than eighty dollars a barrel."

Cheney grunts. "The Russians have been nothing but a source of pain. Gaming us from the start."

Bush seems to deflate as the conversation swirls this way and that. There's no worse threat, that's agreed. A grand global initiative, yes, that's what's needed. But how? At this point, with Bush's low approval ratings and America's diminished global standing, who can lead one?

Soon the hour is up. Bush looks at Rolf, quizzically, time past and time present merging. He realizes, of course, that Rolf is no longer the cowboy operator, launching CIA missions and managing Lovebugs, in the wake of crisis. He's a bureaucrat, head of a business-suited Washington army, offering a "process solution."

From across the table, Rolf sees that Bush has aged and grayed. "You could see the lines on his face, and you knew how he got them," he says later. "We were once immortal. Everything there for the taking, ideas becoming action. That time had passed." He's thinking, of course, of those days after 9/11, when there were candlelight vigils in Tehran and the countries across the planet, even stated opponents, felt a chill. *What would happen if the United States were laid low?* Whatever we did, they would have followed. And at that moment, the United States chose sweeping vision and vengeance, and the world followed, until America seemed to spin in an unexpected direction, toward Iraq, and slip into surprising shadows at places like Abu Ghraib and Guantánamo. Looking back, one could see there were clearly other choices.

As for Bush, his time has passed. The thing that always guided him—his gut, his instinct—would never again be trusted, even by the people closest to him. It wasn't enough—his nerve. It has never been enough.

"Well, Rolf, if there's anything you need me to do," Bush says, "make sure I know."

"Of course, sir. We'll follow up as needed."

A few moments later, he and Bodman are out the door of the White House and into their waiting sedan.

"I feel nauseous," Rolf says, after a long minute. "I feel like I failed. Nothing will come of this, nothing, really."

No, no, Bodman says, reassuring him that it is an admirable starting point, that Bush is now engaged in the search for solutions.

Rolf nods, but it's unconvincing. Talk of solutions makes him wonder why he didn't just shuck off the agreed-upon plan—his negotiated role as bureaucratic representative—and talk of action, of launching a dozen undercover teams to fan out across the global markets, instead of just sitting in the dark pondering hypotheses. But what would be the point of it? No one in the U.S. government can manage that effort, not now. It would just end in frustration for Bush—added frustration besetting a reduced man.

Rolf still wants to send out his teams; he's still committed to it. Maybe Rob will manage it, or maybe he'll leave government and do it himself, Rolf thinks, he and Jim Lawler. In the wake of the big meeting, he tells a friend, "The AT is still my mission."

"AT" stands for Armageddon Test. That's the term he uses sometimes with Rob and other confidants. It came up in early conversations a year ago, and then it stuck. He hated its religious overtones at first, but more and more, it seems apt. The landscape is, after all, mapped with religious certainties, providing the coherence of moral narratives that define identity and, ever more, action. Ancient stories of heroes—God's chosen heroes—who offered visions of an improved mankind, of perfection.

Can anyone doubt that in the delicate balance of faith and reason—the dance of Avicenna and al-Bukhari between faith's perfection and reason's best efforts—the will to power has created the next generation of hungry opportunists? Rolf, like others in the depleted Situation Room of September 2007, can't help but quietly marvel at the ingenuity of bin Laden and Zawahiri—who both now live in caves and who make claims of purity through self-denial that are deeply appealing to so many of the world's people, impoverished and powerless. It is from those unfindable hideouts, with their regular transmissions, that they've harnessed the frustration with reason's failures—so clear now for yearning multitudes—to revive the dark genie of Holy War.

At this point, the great race—what may be humanity's last great race—is already well along, as terrorists race frantically around the globe to get their hands on the crystals in reason's vault; empiricism's diamond, the secret of how the subatomic world really works and the way it, and everything, can be destroyed. That would be reason's dark genie.

The Armageddon Test, after all, is so much larger than Rolf or Rob or their "shock to the system" teams. Everyone, everywhere, is being tested, as never before. In the days after the White House meeting, it's decided by Bodman and others that Rolf should be more public—why not?—and he sits down with *The Washington Post*'s David Ignatius, who writes a predictably startling column about what the "man who thinks the unthinkable" has been saying inside the White House for years.

To mark the September 11 anniversary, Thomas Kean and Lee Hamilton, the famously bipartisan brokers who headed the 9/11 Commission, write their own column, entitled "Are We Safer Today?," asking how, six years later, America can "still lack a sense of urgency in the face of grave danger" and how it can be "possible that the threats remain so dire."

They run down the predictable list: the government's "lack of focus and a resilient foe"; the "rising tide of radicalization and rage in the Muslim world—a trend to which our own actions have contributed"; and the enduring threat housed among "young Muslims with no jobs and no hope, who are angry with their own governments and increasingly see the United States as an enemy of Islam."

After listing the many ways America's reliance on the traditional use of force—and, often, morality-crushing means—to achieve its goals has exacerbated Muslim anger, the duo hits upon their central point:

"Beyond all our problems in the Muslim world, we must not neglect the most dangerous threat of all. The 9/11 commission urged a 'maximum effort' to prevent the nightmare scenario: a nuclear weapon in the hands of terrorists. The recent National Intelligence Estimate says that al-Qaeda will continue to try to acquire weapons of mass destruction and that it would not hesitate to use them. But our response to the threat of nuclear terrorism has been lip service and little action. The fiscal 2008 budget request for programs to control nuclear warheads, materials and expertise is a 15 percent real cut from the levels two years ago. We are in dire need of leadership, resources and sustained diplomacy to secure the world's loose nuclear materials. President Bush needs to knock heads and force action.

"Military power is essential to our security, but if the only tool is a hammer, pretty soon every problem looks like a nail. We must use all the tools of U.S. power—including foreign aid, educational assistance and vigorous public diplomacy that emphasizes scholarship, libraries and exchange programs—to shape a Middle East and a Muslim world that are less hostile to our interests and values. America's long-term security relies on being viewed not as a threat but as a source of opportunity and hope."

Why their column, like that of the graybeards that started the year, gets highlighted and saved in miscellaneous folders far and wide is that, one by one, everyone—from Wendy Chamberlin to the confessional Alan Foley, from Richard Dearlove to Stephen Abraham—is joining the battle to revive hope, reasonable hope, shared and generous hope; to light a path to mankind's best instincts, to the consistent and often inconvenient humility and generosity—the beating heart of moral energy—that has saved us in the past.

That's the real Armageddon Test. And if you're not in on it, you can't possibly pass it.

Because the genies, those pitiless bastards, have spotted each other. And they're counting on us not figuring out a way to get along.

Certainly not in time.

ACT III

The Human Solution

It rains sometimes on Ground Zero.

Then there are days when it's sunny, and people just walk by as though nothing really happened here. And the snow, when it comes, piles up in the deep, wide hole, the foundations of vanished buildings, and it looks lovely, really, coating everything in white.

There's no memorial, but there will be soon. This is the sixth anniversary of 9/11, and before the next one, the scar in the earth will be replaced by an agreed-upon set of symbols, which can't hardly be constructed until grief and memory settle up.

That's been happening. It always does. People resist, but other events, big and small, occur as though the world was invented each morning and retired each night, without a care for what's been, for the suffering and the shattering dread, and the way people held each other.

There's a light drizzle at Ground Zero this morning. And that seems fine. The rain touches everything, equalizes it, moistening the crowd and the tall buildings and the dignitaries, all together.

There's a gathering of about three thousand, many of them families of those who worked in the towers or the firemen who were lost. They carry little shrines, with pictures on cardboard, and maybe a bow or a flower. And no one puts up the umbrellas. They're going to brave it. That's the first response. So they get wet. Just think what occurred here, what's a little water. Some people take out handkerchiefs to wipe water from their faces. Others look up, like they're in a conversation with the sky. Let it come. Mayor Michael Bloomberg is up front, handling things—as he does each year—with nonpartisan dignity. There are politicians in the crowd, including presidential candidates Hillary Clinton and Rudy Giuliani, but they just mix into the crowd, inconspicuous. Bloomberg offers remarks, a small band plays, a singer sings, and, like every year, the names of the fallen are read.

As that occurs, name after name, the heavens open and it begins to pour. No one budges for a minute. No one wants to be first. Someone is, though, one umbrella and a second, then others join in—no point in sitting at some desk all day soaking wet—until there are hundreds of umbrellas popping, spring-loaded and of many colors, in a widening canopy.

It's been six years. Stay dry.

In countless ways, America is healing and moving on, thinking about what's next. There's an urge to solve problems that have deepened while the country drifted through daydreams of revenge and redemption, a pent-up desire to finally see things as they are and to get down to work.

How that occurs will determine a great deal about the country's emerging character, about its confidence in its own capacities, about its faith in possibility.

The challenges on both accounts will be significant, especially in that an indisputable legacy of the September 11 attacks is proof of how meaningless borders are in a shrunken, connected world.

All but discernible, just beyond the horizon line of southern Manhattan, are struggles in countries like Iran, North Korea, Iraq, Afghanistan, and Pakistan that once barely registered a notice. Characters from those countries' unfolding dramas now populate the American imagination and seem to sit as spectators on each year's commemoration at this site. What is it they think when they nose in, often uninvited? Does it matter? Do we need to project a particular image to them, of forcefulness or virtuousness—not just on this day, but on all days?

It's exhausting to feel that pressure, and profoundly liberating to feel it pass, year by year, as the country slowly leaves behind the trauma of being a target and begins, again, to live in the moment, the American moment—an earnest, headlong, bumptious present tense—and invite the world to do the same.

Of course, it's about people in motion, and how they feel, and this tension between past and present is what Ray Kelly feels as he slips away from the 9/11 ceremonies and into his waiting sedan. This is the busiest day of each year for the New York City police commissioner, and before all the names have been read, he's rushing back to his office for a security briefing. Kelly, raised in New York—his father, a milkman; his mother, a Macy's clerk—oversees a force of thirty-six thousand, a small army that has been on high alert since the pre-dawn hours. The city, he feels, is a target on days like this, laden with symbolism, just as it is on the Fourth of July, or Christmas Eve.

"We try to do what we can, but a city this size can't really be protected," he says as the car weaves through downtown traffic. "The city is full of symbolic and iconic things. We guard them all. The key, though, is to think bigger than a famous building or a particular day.

What do these things represent, what's the idea behind the symbol. That's the thing you've got to protect. You do that, you're fine, no matter what."

The briefing with his two top aides—David Cohen, a former head of operations at CIA, and Rich Falkenrath, until 2004 the deputy chief Homeland Security adviser to Bush—covers the day's reports thus far, issues of preparedness and an Islamic cell in New Jersey that they're following.

Kelly, leaving for a lunch with the families of police officers who died on 9/11, walks by a bank of monitors and looks for a moment at the screen for Al Jazeera. There's a video of a statement by Zawahiri. In one recent month, July, al Qaeda's released more than 450 statements, books, records, short videos, magazines, and assorted online products, flooding the airwaves and the Internet. He watches the words crawl across the bottom of the screen. "Deeds," he says. "It's deeds that'll matter. In the long run, it's not what you say, it's what you do. And this is going to be a long race, the one we're in."

For anyone wanting to measure progress in that long race, there happens to be an ideal spot just across the Hudson. Jersey City, New Jersey, is a tough town of a quarter million, brimming with a mix of people from across the world. The town's about a third white, a third black, and a third of other assorted races, but skin color is not the key dividing line in this New Jersey city. A wide swath of the population is from somewhere across the oceans, or their parents are. The streets move with a global energy, many languages and flavors—Egyptian, Chinese, Pakistani, Kenyan—all trying to test American promise at a time of peril.

In an apartment here, Ramzi Yousef, Khalid Sheik Mohammad's nephew, designed the bomb for the 1993 World Trade Center attack. Omar Abdel Rahman, the fiery, blind sheik from Egypt, an early partner and guide to both Zawahiri and bin Laden, preached at a mosque here. Rahman was a spiritual leader of a movement that has since spread across the globe, fueled largely by the way it twisted Islamic concepts into an attack on America and the West.

But as life resumes on September 12, worthy adversaries to Rahman, and his followers, are gathered a few blocks from where he preached.

They are from forty-six countries. They are young, many of them immigrants to America, and most of them poor. They are gathered in a hall, a large hall, listening to a speaker or two and several pieces played on a grand piano.

This group—two thousand strong—are the incoming students at New Jersey City University, a

THE WAY OF THE WORLD · 397

ten-thousand-student commuter college that serves
education to any paying customer in the city, or any
surrounding area.

A music professor plays Paganini. A graduate sings
the national anthem, and the students, many dressed
formally, mix in a sunny courtyard—a girl whose family fled religious persecution in Egypt, a boy who travels an hour each morning from south of Newark on the
commuter train, a Korean kid whose mother washes
floors in the office towers of Manhattan. They drink
the free soda. The sun is out. They talk not about ideology, or evil, ancient feuds, or historical wrongs that
need to be made right. They just talk about being
present—present in their American moment—about
what classes they've planned for the coming semester,
their night jobs, where they've gotten deals on used
books.

Today forces loosed by an army of fiery clerics are
cresting in the world's most dangerous nexus—the
crossroads of nuclear-armed Pakistan and poverty-
crushed Afghanistan. It's on this terrain that an array
of our actors, especially those who've embraced Islam,
race forward to answer questions of both heart and
mind, and test whether America's presence and ideals
are of any value—any sustenance—in their search.

Kelly's police boats, meanwhile, patrol New York

Harbor, as they do every day, keeping a wary eye on its great monument. Of course, there's no way to guarantee protection, not even of the statue. But it's not the symbol itself that matters most. Standing behind the lady are the kids in this courtyard, and beyond them, those slowly awakening across a wide country to the complex demands of moral purpose.

It is they who must provide the light.

Chapter 1
Everything Connects

The Russell Senate Office Building is just ahead, a hundred yards' walk, and behind it, the Capitol, the sun gleaming off its great white dome.

Usman Khosa, after a year of tumult, has recovered his balance. By today—September 25, 2007—he's resumed his full-bodied American sprint.

Which is why he feels he can take a morning off and do something for himself. He's earned it.

He passes through the metal detector at Russell's North Entrance and walks up the marble stairs to the Caucus Room. It is one of the most majestic chambers in Washington, its Beaux-Arts style carrying touches of the classical revival so prevalent in the West when this, the Senate's first office building, was constructed in 1906, the start of the American Century. Usman looks for a seat. The room—site of historic hearings, from Teapot Dome to Watergate to Clarence Thomas's confirmation—is jammed and buzzing.

Usman spots Benazir Bhutto, across the room, at a table near the dais. He's never seen her up close like this. She *is* beautiful, he thinks, the wide dark eyes, the full lips, the perfect symmetry of eyebrows, nose, and chin, like you see on actresses. He knows all the stories—how she may have had her brother killed, how she, or her husband, stole hundreds of millions from the Pakistani Treasury—but he's always identified with her journey, the way she left Pakistan to see the world, to get an education at Radcliffe and Oxford, and then returned, ultimately to lead her country.

He's been thinking, of late, about that last part, about returning to Pakistan, maybe for a short time, maybe longer. The country seems to be slipping into crisis, and he's begun to feel differently about his identity as a Pakistani. It was something he fled from in the months after his interrogation, unable to reconcile his emotions about being singled out, a racial suspect, for something he had always been proud of—his nationality, his heritage. Maybe it loosened his grip, a little, on his Pakistani pride. He drank hard, partied hard, lived hard, like Americans are supposed to do, only to be stopped cold—like someone hoisted in a trap—that night with his friend Pervez, the strip club, and the visitors from

Pakistan who saw a path to heaven in killing him. The Muslim catch, the war raging inside this religion of his. There was nowhere to escape it. That he now knew. After all the whipsawing, he knew he wasn't going to stop being who he was—a Muslim. He needed to find resolution, a balance between his faith and his ambitions.

A pretty, petite girl slips into the chair next to his. "I can't stay long. I have a class at noon," she says. She's a Georgetown sophomore of Pakistani descent with the same name as Usman's sister, Sadia. They've known each other almost a year, and they've been dating seriously since the spring.

"She's gotten old," Sadia says, craning her neck to see Bhutto.

"Yeah, I suppose," Usman says. "I wouldn't mind talking economics with her. A lot of her programs seem sort of socialist to me."

"You're such a capitalist pig, Usman."

He smiles. He really likes Sadia—she's smart and funny and challenges his notions about, among other things, how to be a Muslim in America. Her family is religious, but wildly engaged in secular achievement. She was born here, went to a strict Islamic school near Chicago, and seems utterly comfortable in her own skin. Her father was the immigrant, like Usman, and rose steadily from bellhop all the way to owner of one of the first Muslim-run hotels in the United States. For Usman, this represents an archetypal American journey—maybe similar, in ways, to his own. Of course, Sadia's father did all that in an earlier time, before wars inside of Islam started to explode across the globe.

Now Bhutto is preparing to step into the middle of that war and bring her complex profile to the battle over how to define a religion. "This should be interesting," Usman says, nudging Sadia. "Let's see what she's got."

Fifty feet away, Wendy Chamberlin steps up to a microphone and looks across the room with delight. Everyone's here. Former ambassadors, top State Department officials, heads of think tanks, national security types, and the media, both electronic and print.

It was a coup scheduling Bhutto, something Wendy nailed down a month ago, making the Middle East Institute the sponsor of today's speech. But that was before Bhutto announced just two weeks ago that she planned to return to Pakistan on October 18 and participate in fair elections at the soonest opportunity, setting up a real moment of global drama.

When Wendy steps to the microphone to introduce Bhutto, she's instantly stymied by the need to be utterly neutral and says that while the "Institute doesn't take policy positions, it is a strong backer of the democratic process in Pakistan." She mentions—for the second time in two minutes—the institute's website, thanks Bhutto for coming, and steps aside.

Bhutto then steps up and gives a half-hour stump speech—a political speech—mostly about Musharraf's significant deficits, her fierce opposition to extremism, and her readiness to be the next leader of Pakistan.

Wendy watches from the wings, feeling deflated. It's as if Bhutto is auditioning for American support, advertising how she'll be tougher opposing radicals than anyone else on the landscape—like democracy's winged avenger. She says nothing about the real Pakistan, about its people and their real challenges, such as poverty and illiteracy. Wendy hopes for better during the half hour of Q&A, the questions written on cards and passed up the aisles.

Holding a microphone, Wendy reads them one after another—a question about how Bhutto will control the country's powerful military, another about how she'll tame insurgents in the tribal regions, a third about whether she'll give the West access to A. Q. Khan, something Musharraf has blocked. Her answers are complete but predictable, and often as carefully hedged as her speech, including an assurance that she'd allow Khan to be questioned, but only by the International Atomic Energy Agency.

That is, once she's prime minister. Of course, that's part of the fiction that dominates the room, as Washington's policy community—largely ignored by the Bush administration for six years—pretends, in this august setting with its frescoes and Corinthian pilasters, that it still plays a role, any role, in helping the government stay on some wise course, and that America still retains anything close to the clout needed to manage Bhutto's reentry and bring democracy to Pakistan.

As Bhutto goes on about her desire that the United States will, at some point, act as an election monitor—something America once had the honest broker's credibility to manage—Wendy finds her eyes wandering to the strange cap etched in the crystal globe of one of the tiered chandeliers. She's quite sure it's called the Liberty Cap—the brimless hat of freed slaves in the Roman Republic and a symbol in the American Revolution of opposition to tyranny. What question would a clutch of Founding Fathers ask Bhutto, or what path, more important, would they chart to help America restore its moral authority, Wendy wonders, as she pulls up the next card and reads it. It's a question from someone in the Netherlands embassy, about how Bhutto "could possibly remain her own woman in a coalition with Musharraf."

The question touches upon both Bhutto's character and the deeper issue of what the United States actually has the power, or will, to force Musharraf to do. Three hundred people seem to lean forward in their seats.

Bhutto pauses. "I would like to remain my own person," she says, in a different voice, tentative, almost apologetic, and describes how she was "very much my own person" in her first term, but then "decided to be co-opted" in her second term in order to return to power. "But I think on the third time around, being over fifty now, I would like to be my own person, even if it means not lasting very long. Because even when you try to last long by making the compromises— compromises that you think will help you stay in power—you still don't stay long anyway."

Wendy, looking on with a startled expression, is not sure she heard Bhutto correctly, and then she knows she did and that this response—humble and honest— might actually be the start of something, an unexpected evolution.

Wendy Chamberlin suddenly feels alone in this crowded room, just she and Bhutto.

"Thank you," she whispers into her mike. "Thank you for saying that."

406 • RON SUSKIND

A half hour later, Bhutto is in the middle of a warm, noisy scrum. Pakistani journalists, think tankers, well-wishers—all wanting to get close. She works through them, one at a time. She'd been to Capitol Hill many times during these last eight years of exile. There were never crowds like this. Now she doesn't want it to end.

It's down to a dozen diehards when a well-dressed young man steps forward.

"Madame Bhutto," he says. "My name is Usman Khosa."

"Usman, very nice to meet you," she says. "You're from Pakistan, I gather."

"Yes, my family's from Lahore. I went to the Aitchison School, and then I came here, to Connecticut College." He feels surprisingly comfortable with her—her face is gentle and unimposing up close—and he talks a moment about his life in Washington, about how he works as "an economic consultant and someday maybe, I'm hoping, I'll go to Yale Management School."

Usman stops—what's he mentioning Yale for? *Like he wants her to write a recommendation.*

Bhutto smiles. Something's dawned on her. She looks at him intently. "You know, part of what I hope is that I can change our country so someone like you will want to return and bring your talents and your education back to Pakistan. We will need you."

He nods, not wanting to let on how she's just hit a bell inside of him. "Yes," he says, his voice trailing off. "I'm thinking of that, too."

Once the crowd is gone, Benazir Bhutto turns to Asif Zardari, her husband, and asks girlishly, "Did you think it went well?"

He smiles and nods noncommittally. "Fine," he says. "Let's keep moving." And they do, Bhutto and her small entourage, down the hall from the Caucus Room and quickly into the office of Massachusetts senator John Kerry—a senior member on the Foreign Relations committee.

The priority of this trip is to get Bhutto the security support she lacks. October 18 is only three weeks away.

They all crowd into Kerry's office. Bhutto, parched from hours of nonstop talking, wonders if she could have a drink, and a hubbub ensues: Kerry's office lent out their crystal pitcher and glasses to the Middle East Institute for the event; a Kerry staffer runs across the hall to borrow Ted Kennedy's set.

Once everyone is refreshed, Kerry is swift off the

408 • RON SUSKIND

mark: "This is a volatile situation you're walking into, Benazir."

The United States, he says, is generally hesitant to ensure the protection of anyone who is not a designated leader, a provision to prevent U.S. forces from becoming embroiled in the internal disputes of sovereign nations.

"Senator Kerry, I want Pakistan to provide me with the security I am entitled to under the laws of my country. I'd be grateful if you would talk to the Musharraf government and tell him the U.S. expects he will fulfill those obligations."

Kerry sighs. Of course, he, a senator, can't conduct unilateral foreign policy. "Well, Benazir, I will certainly talk to the State Department about that point being made to Musharraf," he says as forcefully as credulity will allow.

There's not much else to say. They chat about people they know in common, and there are plenty. Bhutto has a community of what she calls "dear friends" that stretches from U.S. officials in the administrations of the first president Bush and President Clinton—whose years in office overlapped with her two terms as prime minister—to old friends from Harvard, many now prominent thinkers and writers, to assorted notables, such as Ted Kennedy, across the hall. Her current fortunes, however, are in the hands of a half-dozen people beyond her orbit: a tight circle of policy makers in senior posts at the State Department and in the Vice President's Office. All official contacts with Pakistan on Bhutto's behalf must be channeled through this small group, overseen, in essence, by Cheney and Rice, a duo with a long history of internecine combat, most of it dominated by the vice president.

Rice's latest goal, befitting her role as secretary of state, is to restore a modicum of energy and credibility to U.S. diplomacy. For five years, U.S. foreign policy—American diplomacy—has consisted of little more than the assertion of Washington's will on both friend and foe, with penalties for noncompliance. Brokering a Musharraf-Bhutto partnership would be a change from all that, facilitating a marriage that could, with eventual elections, let some of the steam out of a country ready to explode. If that were to happen, it would offer proof, finally, of a central plank of Bush's rhetoric: democracy is the best response to extremism.

Rice's assistant secretary of state, Richard Boucher, is managing the Bhutto return, but only under the wary gaze of the vice president. Cheney, like Bush, is against abandoning Musharraf. Both feel he's made some errors, especially in signing a peace deal in September 2006 with the Taliban and assorted radicals in the tribal areas, but they prefer him to Nawaz Sharif, another former prime minister, who's also angling to return from exile and has preserved close ties to Pakistan's growing Islamist movement.

For the United States, the driving concept is that Bhutto will save Musharraf, giving him the electoral credibility he lacks and giving the United States more leverage in making him the man they've long wanted him to be: an obedient warrior against the Taliban and al Qaeda.

That last bit is the real goal for the White House, and it is becoming a more urgent matter by the day. Pakistan is, after all, the epicenter of global concerns— with fifty-five nuclear weapons, an abundance of fissile materials, vast poverty, and radical movements that are large, growing, and deeply entrenched in both the military and intelligence services. There could scarcely be a worse confluence in a single state, much less one that is now relied upon to confront the world's militant Islamic surge, which has largely settled within its borders. Musharraf brokered his 2006 peace initiative with what was essentially a newfangled nation-state, calling itself the "Islamic Emirate of Waziristan." That area, comprising the southern half of the tribal areas, is most likely where bin Laden and Zawahiri are now hiding, and where, according to U.S. intelligence, al Qaeda has reconstituted itself into a core group of fifteen hundred or so operatives and managers.

What has been taking shape over the past few months, right up to today, is the most revealing of choices. Rhetoric aside, what does the United States really trust in terms of confronting extremism: raw force wielded by a military dictator, or the democratic ideal it espoused to justify actions in Iraq, Afghanistan, and much of the globe?

Bhutto is clearly betting on the latter. "Democracy is the only response to extremism" was her signature line in the morning's speech. She's sure the United States is behind her; Condi and Boucher have told her as much. Her faith in them exists because she detects comforting self-interest: they want an advertisement for how democracy can work, whatever the underlying reality. In her mind, the appearance of a step forward for democracy will be plenty, as long as she can return to power.

Her problem is that she knows Musharraf better than anyone in the United States. He was a rising commander in the Pakistani army in her second term as prime minister, and she feels he'll accede to this power-sharing deal only under duress. He must be forced into it by the United States.

Boucher has told her not to worry, that Musharraf's on board. But he says she needs to work out her complex issues with the general and negotiate a win-win deal. Bhutto's demands are not insubstantial and hinge on being granted amnesty from long-standing corruption charges and being made eligible for a third term as prime minister, something forbidden under Pakistani law. But Bhutto holds a trump card in her hand: legitimacy. If Musharraf accepts her demands and agrees to step down as army chief, she'll stop her party, the PPP, from resigning from the national and provincial assemblies, as other opposition parties are threatening to do. With the large and populist PPP voting in the assemblies, Musharraf's reelection as president on October 6—a parliamentary plebiscite by his ruling party—will appear far less illegitimate. Then following the election, Musharraf would "take off his uniform" and Bhutto would return triumphantly to lead her party, the country's largest, as the de facto prime minister until the parliamentary elections install her in the actual post. Work all this out, Boucher has told Bhutto, and Musharraf has assured the United States "he'll keep you safe."

As they walk the halls of the Russell Building, Team Bhutto talks darkly of such assurances. They are, by nature, a suspicious lot, and fiercely protective of their captain. For no particularly good reason, Bhutto is surrounded by Arkansans, led today by Larry Wallace, a silver-haired Benton lawyer who's been close to Bhutto for three decades. Mark Siegel, Bhutto's K Street lawyer and fixer, leads the Washington corps, backed up by the pert brunette public relations guru Lisa Cotter Colangelo. In a wide-lapelled pin-striped suit, Zardari—looking like an extra in the Karachi road show of *Guys and Dolls*—walks astride Bhutto as she makes a grand entrance into the Russell Building's private dining room, mostly for senators and senior staff, her entourage in tow.

The group's host is Arkansas Democratic senator Mark Pryor's chief of staff, a guy named Bob Russell, who seems to nod—"Yes, that's her"—toward a roomful of craning necks as Bhutto and her entourage settle at a central table.

Before the water glasses are filled, the discussion is well along about the security concerns and how Bhutto can direct some added pressure toward Musharraf. Larry says he's talking to private security contractors, and he has someone named "George" from Israeli intelligence helping out. There's some jaunty talk about "jammers," which emit a low-power radio frequency that either detonates radio-controlled bombs or blocks their ability to receive a remote signal. Russell says he can get about fifty Senate chiefs of staff to look into Bhutto's security dilemma—they're meeting in Philadelphia in a few days; a steady stream of staff chiefs have already come by the table to pay their respects to her. She thanks them in advance for their help, as Zardari—from across the table—says, loudly, "I spent eight years in prison while she fought for my release. Now it's her turn to go to prison and I'll be the one who gets famous by fighting for her."

"Yes, Asif, it's my turn now," Bhutto says, happy, it seems, that Zardari—who speaks in the blunt macho patois of a South Asian tribal chief—got a laugh with his line.

"I'm so enjoying this lunch, and it's dinnertime for me in Dubai," she says cheerily to no one in particular, and it's clear that the logistical talk of security is giving her a sense of immediacy about the planned triumphant return. Finally, after all these years, it's really happening, just weeks away, causing a global frisson and placing her where she ought to be, at the center of attention.

"I know these security concerns are there, but I still feel optimistic, very optimistic," she says to the table. "I know there are roadside bombers and suicide bombers. But I feel the crowds and everything will make it difficult for the bombers to get close. If we can get jammers to jam them that will be very helpful."

Bhutto then says she wants to go to the U.S. Senate

gift shop to buy things for "the kids"—"I love so many of the items there"—and everyone's up and moving, out of the dining room and into a long, majestic hall, trailing behind Bhutto and her white headscarf.

Zardari races to catch up to her. They are, of course, an unlikely couple, and the subject of tireless gossip in South Asia. Who knows what sort of hard deals they've cut over twenty years in a marriage arranged in 1986 by Bhutto's mother because—as she told Benazir—"you must be married to be prime minister." She was right, however the union might have affronted Benazir's emancipated sensibilities, or pained her as Zardari lived the life of a crime boss—called "Mr. Ten Percent" in Pakistan—staying busy, allegedly, with backroom deals. He stays mostly in New York, while she lives in London and Dubai. But they are fused, now, in this storied return to Pakistan. Both of them want to go home.

"This is a beautiful building, it's a beautiful town," Bhutto says as she walks into the Russell Building's cavernous formal lobby, Zardari beside her. "I can't help but feel confident. This is the capital of the world. And if they're behind me, everything should work out. It really should."

As Zardari hears this, he seems struck by a liberating thought. "I believe in strong women. They cannot be controlled. They will outlive us!"

Suddenly the couple turns. One of Bhutto's aides is rushing toward them, saying he's just gotten a call from one of Musharraf's aides. The aide says that Musharraf can't support Bhutto on a key demand—the repeal of the provision prohibiting a third term for prime ministers—and that he wants to talk to her.

An hour later, Mark Siegel escorts Bhutto and Zardari into the office of California's Democratic congressman Tom Lantos for some privacy. Bhutto takes the call from Islamabad.

"The twice-elected provision is important to me," she tells Musharraf. "If you're retreating from that, what can you give me? Maybe some real reform in the election commission?"

He says she shouldn't be hoping for much there, either. In their many calls, he's been surprisingly cordial, often quite reasonable. But something has changed. His voice is harsh, almost mocking her.

She asks if the U.S. officials have had "conversations with you that make it clear that my safety is your responsibility."

Yes, some have called, Musharraf says, and then laughs. "The Americans can call all they want with their suggestions about you and me, let them," he tells her. His change of tone, she doesn't realize, arises from those very talks with American leaders in which she's invested hope. Musharraf has been in a running conversation with Bush and Cheney since the two men arrived in their high offices, right up until a few days ago. He knows that when confronted with a choice between power and principle, they'll go with power, not even recognizing the sway of certain principles that they loosely espouse. Democracy, accountability, justice? Americans won't be backing up any of those ideals with force and consistency, Musharraf knows. Not in this era. Such ideals are just means, when convenient, not ends. Bush and Cheney have made that perfectly clear. All things being equal, they'd prefer someone freed from those lofty principles to face down the angry Islamist threat. What Musharraf can't believe is that Bhutto—a very tough strategist, deep down—doesn't get this. That's why he's laughing in her face.

He finishes the call with a dose of fair warning.

"You should understand something," Pervez Musharraf says, finally, to Benazir Bhutto. "Your security is based on the state of *our* relationship."

She hangs up the phone feeling as though she might be sick.

On this very day, Ibrahim Frotan also thinks hard about a veiled threat.

He is back in Bamiyan, holding a letter that was left

on his doorstep this morning. It is unsigned, but clearly from some of the religious fundamentalists in his town. It reads, "We know you've been to America," that "you have tasted of their corrupt ways," and "we are watching you."

He's afraid, and not sure whom to turn to. He wonders if he should go to the Bamiyan office of the PRT—one of the Provisional Reconstruction Teams stationed across the country to deal directly with the Afghan population.

It's not the kind of place where you just stop by. Most people in town have never been. Talking to an American military officer represents, still, the crossing of a line between us and them, about the last thing you'd want to do.

But there's a twist. He puts down the letter and digs beneath some clothes in his dresser for a business card, packed in one of the small troves of items—well hidden—from his time in America. The card was given to him by a man he chatted with at Niagara Falls, who said he had a friend in the U.S. Army stationed in Bamiyan.

Ibrahim and the man, an older fellow, agreed that this was an amazing coincidence. But, as he looks at the laminated card and runs his long fingers along its edges, Ibrahim feels that maybe there are no coincidences in this world. That everything is connected, somehow. Things happen for reasons, and good things are meant to happen once in a while for him, for a young man named Ibrahim Frotan.

That such a swell of confidence should come from a business card—or from any of the other items in his drawer of talismanic artifacts—is testimony to Ibrahim's time in America, especially what happened after he left Denver. He sent, after all, not home but rather to Pennsylvania, and it was his family there who took him to Niagara Falls, just before he returned in the late spring to Afghanistan.

To understand how the world really works, when it does work, you have to step into Ibrahim's shoes, the ones bought by Ann Petrila, as the Muslim boy travels, eyes wide open, from Colorado to the very dreariest corner of Pennsylvania.

It wasn't supposed to happen. Ibrahim was right. They were ready to send him home. At the airport in Denver, he spoke in an urgent whisper to the ticket agent: "Please tell me where this ticket is for." She looked at ticket, then at Ibrahim's worried face. "It's for Alabama," she said, opening the ticket for him. "Says it right here." Ibrahim nodded. "Okay, but does it connect to Afghanistan?" She said no, and he boarded a plane for Mobile and the home of an older man, a friend of the American Councils, who put him up for two weeks while Naeem Muhsiny tried to figure out what to do. The resistance to sending Ibrahim home came from the organization's Afghanistan headquarters in Kabul. Bamiyan had just elected a woman governor who wanted to begin placing girls—maybe lots of them—in the exchange program. If Ibrahim were sent back, the head of the Kabul office complained, "It could blow the whole thing."

Fine, but where to send him? There was no school willing to accept him in Alabama. So Naeem sent out emergency notices to American Councils coordinators across the country.

Soon the alert reached Kane, Pennsylvania, and the home of Mary Lisa Gustafson, a French teacher, and her husband, Tom, who works in a lumber mill.

It's a second marriage for both of them. Their children are grown, and Mary Lisa sort of runs the show, much like Ann Petrila. The two women are actually quite similar. About the same age, early fifties, with husbands who are away a lot. The lumber mill where Tom works is not always operating at full capacity, so he does odd jobs to make ends meet. Mary Lisa teaches French at a school down the road, in Johnsonburg, a town shadowed by a giant paper mill that gives off an impressive odor. There are no complaints, though. At least, in Johnsonburg, there are jobs.

Most of the rest of the industrial towns in this north-western part of the state, including Kane, are not so fortunate. Factories have fled, one after another, for cheap labor overseas. Unemployment here is more than 25 percent, if you count the people no longer looking for jobs—the result of twenty years of downward drift.

Like Kane, Mary Lisa hasn't had the easiest two decades. She's from around here, from Erie, took French in high school and, in college, matched up with a Moroccan boy—a French speaker and a Muslim—who was the star of the school's soccer team. That'd be her first husband, long gone, who now lives in Florida. Her two sons, whom she raised, are living in upstate New York.

When, eight years ago, she met Tom, a quiet mill hand from Kane, gentle and dependable, she said, "Why not?" Which is what Tom said back, when Mary Lisa, in 2003, showed him a flyer she'd picked up at school, about Afghan exchange students. Their house was small, but they had an upstairs bedroom, and it was far from too late for them to try some new things in their life.

In the fall of 2005, the bedroom was home to Fazila,

one of the genuine stars of the American Councils' program. A brilliant girl from Kabul, Fazila soon became fast friends with Mary Lisa. She blazed through her studies. Eventually, she got a full scholarship to Sweet Briar College, in Virginia.

Naeem remembered that Fazila had said only the nicest things about her time in Kane. Calls were made, and Mary Lisa found herself sitting in the office of Kane High School's principal. He remembered they'd had one good experience with a kid from Afghanistan. Why not another?

Soon Ibrahim was walking the halls in a daze. It was the second week in February. A teacher from the school showed him where his classes were—a light schedule, mostly English, art, and phys ed—and then let him wander.

That's where, in the hallway, he was spotted by a girl.

She walked right up, asked who he was, and said, "We should be friends." First day.

Her name was Jillian Davidson, and she was used to taking in strays. She'd grown up in Kane, like almost everyone else around here, and her mother took in foster children—a conventional way for a woman with a strong maternal instinct to make ends meet. In fact, about thirty foster kids, of all shapes and sizes, had passed through Jillian's house by the time she was in elementary school.

Jillian saw Ibrahim's face—a handsome face—and saw the look of someone who was lost. She'd seen quite a bit of that.

Ibrahim was open, at this point, to anything. The experience in Denver had tenderized him, giving him a kind desperate readiness to grab whatever passed by him and hold on tight. Already, looking back, he felt remorse about so much—how Ann and Michael and Ben had tried to make things work amid all the crazy twists and turns, how everyone carried a little blame about the way it turned out, and how there was plenty on his shoulders, too.

That, though, was the past. Now he needed to make things work. This was, after all, one of the most powerful natural forces on the planet: a second chance.

They started to talk, he and Jillian. He realized that the last time he'd spent this much time talking to a girl was when he was a child. In his village, all the kids played together, at least until puberty. Once they crossed that river toward adulthood, the gender divisions of traditional Islam took hold. Since then, he had never really talked to very many girls. Back in Denver, Jasmine—who had a serious boyfriend, and seemed very attentive to the rules Ibrahim was raised with—

had been the first. They used to talk often, but nothing like this. He and Jillian were on the phone almost every night. A thousand bits traveled between them across e-mail and instant messaging.

He told her everything about his life, about growing up in Afghanistan and being Muslim, about his older brothers, long gone, whom he doesn't remember as well as he'd like, and about his sisters—how she reminds him of one of them—and his younger brother and his illness. She listened. No matter what he told her, she wanted to hear it.

In school, they started to learn where to walk so they'd see each other at the change of classes. A passing encounter in late February:

"Hi, Ibrahim."

"Hello, Jilliaaaan."

"You know, you say my name funny. But I sort of like it."

"You think it's funny. Well, that's a good thing."

That day, she went to math, he to English, but he thought of something he wanted to ask her during their time after school, when they would sit together in the cafeteria and wait for the buses to come.

"How do I tell a joke in English?" he asked her that afternoon.

"Okay, Ibrahim," she said, "this is called a knock-

knock joke. It's a basic American joke, but there are a lot of different ones."

This went on for a week, at least.

Ibrahim would try—"Orange? Orange who?"—and Jillian would laugh until tears rolled down her cheeks.

And they were lovely cheeks. Ibrahim couldn't help but notice. Jillian was one of the prettiest girls in the school. He felt he had a very good sense about this. He had seen the heroines in the Jean-Claude Van Damme movies and in those with Stallone and Schwarzenegger. Most of the women were blond—Jillian had brown hair—but her face was like theirs, with a small nose and blue eyes; she had a perfect face and a very nice figure. She was popular—everyone liked her—and a good student, straight A's, but she didn't have that many very close friends and didn't go out with the others kids on weekend nights. Ibrahim knew this because that's when they'd spend whole nights talking on the phone.

On a Saturday evening in early March—now about a month after Ibrahim arrived—they had settled in on the phone. He used the one in the small den off the Gustafsons' kitchen after Tom and Mary Lisa went to bed. He would call Jillian and then she'd go to her bedroom—he had never been to her house—and call him back. During that week, in the lunchroom, he'd had a new request. "Now that I am great with a joke—

which is a joke, because, I'm a terrible joke teller—could you show me how to tell a story?" She said she'd try to think of one.

Once they'd both settled in to phone position, she said she had one—she'd created a story. It was about a boy who landed on an island. His ship had wrecked. There were people on the island but they seemed strange to him, and he went his own way, hiding in the forest. He was an older boy, old enough to survive, but he was lonely, and was sure there were monsters on the island, hiding everywhere.

Ibrahim was rapt. The story went on for a half hour, full of twists, and then a character appeared—a girl—who'd wandered into the woods. The boy had seen very few girls, but she was daring and courageous, and he let her befriend him. Day by day she convinced him that there were no monsters he needed to fear. She guided him out of the woods and into a village, where everyone lived, and the boy and the girl "were friends forever."

She finished and was quiet. "Well, did you like it?"

"Yes, Jillian. It was fabulous story, amazing."

Then something dawned on him.

"Is this a story about you and me?"

He could hear her smiling through the phone. "Maybe."

———————

It was just a few days later that Mary Lisa asked Ibrahim if he knew when his birthday was.

He told her he wasn't sure, but that he was once told that it was in the spring.

"That's close enough, Ibrahim. We're going to have a birthday party for you," she said. She asked if he had friends from school he wanted to invite. He mentioned a few people, including a boy from Afghanistan—a more polished, sophisticated boy—here with another exchange program, and of course, Jillian. Mary Lisa filled out the list with a few kids from the International Club and the family of the regional coordinator for the American Councils.

All of them crowded into her small living room, with the old spinet piano. Mary Lisa brought out a cake as everyone sang "Happy Birthday," including Ibrahim, who'd never had a birthday party and was uncertain about the protocol. There was great laughter about this, from Ibrahim as well—"We sing to *you*, Ibrahim; it's *your* day!" The cake, chocolate with white frosting, was delicious.

Satiated, Ibrahim flopped on the couch as a few people started to leave. Jillian came over to him. She sat next to him and said quickly that there was something she wanted to show him. She reached into her purse, pulled out a photo, and handed it to him.

"I've been wanting to show you this for while," she said.

Ibrahim looked down. It was picture of Jillian and a baby, a little girl.

"That, Ibrahim, is a picture of me and my daughter. She's just over a year old now."

Ibrahim looked at the photo, then up at Jillian, then back down. He didn't think he was breathing.

"She's a very pretty little girl," he said.

He looked at Jillian again, and she looked back. Neither of them said anything for a few seconds, maybe longer, her smile tight across her teeth. Then she wished him a happy birthday and was out the door.

Ibrahim couldn't wait for everyone to leave. And they did, thankfully, in a few minutes.

Tom had gone up to bed, and Mary Lisa was doing the dishes. Ibrahim walked into the kitchen.

"Mom, can I talk to you?"

"Sure," Mary Lisa said, drying her hands with a dish towel as she sat down at the kitchen table, Ibrahim across from her. "What's wrong, Ibrahim?" she said. "Are you not feeling well?"

Ibrahim took a minute, trying to get the English just right.

"I think that Jillian is married."

Mary Lisa looked at him, puzzled. Ibrahim was wide-eyed.

"Mom, she has a baby!"

Ibrahim could see her expression change in an instant as it all clicked for Mary Lisa—why the name Jillian had seemed vaguely familiar, and how, two years before, Fazila had mentioned that a girl in the school, a tenth-grader, had had a baby. Even with the poverty in Kane and plenty of single mothers, that was an event.

Mary Lisa now gathered herself. She wanted to choose her words carefully.

"Ibrahim, the baby's father and Jillian were not married."

Ibrahim said nothing, his lips pressed together in a thin line.

"Do you know of this sort of thing," she said, "I mean, in your country?"

"In Afghanistan," Ibrahim said, "we would have to kill her."

Mary Lisa, a French teacher in a mill town who always had to go her own way, who now lives in just about the most forgotten corner of America, then managed to summon the transforming question of her culture, a land built on the revolutionary idea that the people are the sovereign, the bosses, captains of their own fate.

She said, simply, "But what do *you* think?"

The idea that what this boy thinks actually matters is an affront, of course, to all those who've collected power and wealth and authority, religious and secular, based on telling people what to think and having it stick. Ibrahim knows what he's supposed to think, and if he were still in Afghanistan, part of a stone-throwing mob, what he's supposed to do. As he looked up at the kitchen clock above Mary Lisa's head and heard water draining from the sink, he seemed to grasp the truly disruptive revelation of this age, his age: that people across the world's wide valleys are entitled to think for themselves. What a mess, what a glorious mess. Because their decisions, multiplied endlessly, piled high upon one another, are fast altering the landscape, and resisting clumsy, ham-handed efforts at self-interested, top-down command and control. What's hopeful— what Ibrahim's journey shows—is that people will generally bend toward the sunlight, like all living things, if the cross-border conversations carry enough honesty and humility and patience. There's plenty of that here. In fact, all the participants in the struggle over the heart and mind of Ibrahim Frotan—from Bamiyan to Denver to Kane—are at the kitchen table with Mary Lisa. And with Ibrahim, as he dives down to the bedrock of religious doctrine and spots, momentarily, the hairline crack between censure and compassion.

He looked at Mary Lisa. After a minute or two of silence, he was ready to answer her question about what he thinks of this out-of-wedlock mother who would have been long dead in his hometown.

"But she's my friend," he says softly, simply.

The next day they saw each other in the halls.

Two months later, the Muslim boy from Bamiyan took the teenage mom from Kane to the senior prom.

And, good God, how they danced that night.

Ibrahim thinks of all this, of lovely Jillian, and Mary Lisa, and the trip to Niagara Falls, as he flips over the business card. He reads the name on the back—David Jay, PRT—and puts it back in the drawer, among the other items.

He decides he'll go see the PRT man at a moment when it's safe. He can tell the officer that he met his friend, the nice older man, on a platform overlooking Niagara Falls. Maybe the PRT man can help with the threatening letter. Maybe not. But it's worth making contact.

That's one thing he knows. Making contact with people not like you is one of the best things a person can do.

You never know how it'll all turn out.

Chapter 2
Ezekiel's Sword

The key phrase in Arabic is *tahta tawila*. It means "under the table." There are versions of the phrase in the languages of every Muslim country in the Middle East and South Asia.

It means that nothing is as it appears. Whatever is visible, whatever is stated or arrayed as proof, is irrelevant, intended to conceal what goes on behind the scenes—beneath the table—where everything important occurs.

The fact that nations in this part of the world accept such a dispiriting state of affairs is due largely to what didn't happen here: the seventeenth-century's Enlightenment, a period that saw the first foundational attempts at building concepts of accountability, reason-based analysis, and informed consent into a social contract between the rulers and the ruled.

America has recently tabled some of these principles in favor of a fierce tactical model. Its latest generation of political managers and war-on-terror strategists—masters at working the gaps between what is proclaimed and what is quietly enacted—have striven ardently to restore some old world, kill-or-be-killed practices.

But they're out of their depth. This sort of brutal gamesmanship has never been America's strong suit, and this is painfully clear in the fall of 2007 as U.S. officials move into the final turns of brokering a deal between Musharraf and Bhutto. In this case, the complexities seem to grow with each passing hour, as do the stakes. With Afghanistan to the north, Iran to the southwest, and a reconstituted al Qaeda within its own borders, Pakistan is the linchpin of an entire region shattering in the collision of religious moderates and extremists.

The first bad break for the United States comes in early September: Assistant Secretary of State and South Asia Chief Richard Boucher, who's been handling the Bhutto-Musharraf talks, falls ill and needs to be hospitalized. Condi Rice herself tries to step in. She calls a London hotel where Bhutto is meeting with Pakistani supporters. Bhutto doesn't take the call. "Someone said that Condi Rice was on the phone," she said later, "and I thought they were joking." On the third try, Rice reaches her. She and Bhutto talk several times through a long night and into the next morning, ironing out some sticking points with Musharraf. Bhutto tells Rice she's concerned about her security but doesn't recount the harrowing Capitol Hill conversation with Musharraf. She's suspicious that the United States sees her value mostly as a means to shore up Musharraf— rather than as a champion of democratic ideals—and to describe her exchange with the general would show just how untenable a couple they'd make. Musharraf, meanwhile, accedes to several of Bhutto's demands and assures the State Department that her security will not be a problem. Though he commits thousands of troops and police officers to patrol Karachi for her return, it is committed intelligence work or a back-channel deal with the country's radicals—not troop strength—that's needed to stop an attack. That he won't do.

Two days before she boards the plane, Bhutto is concerned. Her team has been frantically trying to beef up her security. Everyone has his responsibility. Her husband, Asif, believes in bodies—surrounding Benazir with as many Pakistani security personnel as possible. He has several hundred lined up to meet her. The rest of the team doesn't think that will be adequate. Mark Siegel and Larry Wallace, Bhutto's American advisers, have been working the problem with Blackwater.

In September, representatives from the firm flew to meet with Bhutto at her home in Dubai and laid out several security plans, each costing about $400,000 per month. They intended to work in conjunction with affiliated firms inside of Pakistan, because Musharraf had blocked visas from being issued to imported American security personnel for Bhutto. Bhutto herself is anxious about being ushered into the country with security forces from America and connecting herself to Blackwater. She turns the firm down. She knows that the United States has accepted Musharraf's assurances that he had her security under control, but she doesn't trust him and sends the general an "in the event of my death" note, identifying various hardline Islamist officials in his orbit who should be held responsible in the event that she is killed.

Bhutto arrives in Karachi on October 18 and begins a long procession to the tomb of Muhammad Ali Jinnah, the founding father of Pakistan, who imagined a state where religion and individual rights would stand side by side. There she will pray. Karachi, a port city of 12 million on Pakistan's south coast, is the base of Bhutto's support, and hundreds of thousands show up for her return. The motorcade, with Bhutto on a platform atop a kind of open-air bus, slows to a crawl in the thick mass of supporters. It takes ten hours to go less than ten miles. Two hundred PPP security officers ring the bus as it moves slowly forward. The jammers given to her by the government—a hollow show of will—don't work.

The pace of the procession has thrown off the timing of Bhutto's arrival at Jinnah's tomb. She's hours behind schedule. By odd coincidence, however, as her ensemble nears the tomb, the streetlights go black. It's after midnight, and her security personnel strain to see into the darkness. Crowds throng in close. Bhutto, who is due to give a triumphal speech after the ceremony at the tomb, steps down from the platform and to the back of the bus to work on the speech's finishing touches. She's been on her feet for hours. She takes off her shoes.

Outside, a man approaches with a baby for Bhutto to kiss. Seeing she's left the platform, he moves to the car just in front of her bus. Security tries to wave him off as he moves back toward Bhutto's vehicle. Someone yells that it's not a baby he's holding.

The blasts, in quick succession, rock the street, as two suicide bombers detonate and dissolve, each man's head flying a hundred yards. Bhutto happens to be behind a partition in the back of the bus. As she flinches, blood and body parts splatter against the Plexiglas. Mayhem ensues. Then it's just darkness and screams.

It is the worst suicide bombing in the country's history: 140 dead, more than 450 injured. The dead include 50 members of Bhutto's security team and 20 government police officers. Save for the 2 bombers, almost all the other victims are supporters of Bhutto and the PPP.

By the next day, Musharraf calls Bhutto at her estate near Karachi. She accepts his sympathies reluctantly.

"I'm not the enemy, Bibi."

She says little. She knows the lines are tapped. It's a new hand, and she's not showing any of her cards.

He's gonna die," Candace Gorman says. "He's just so sick."

The two aides to Illinois congresswoman Jan Schakowsky gaze on, solemnly.

They listen intently for the next fifty minutes as Candace tells the story of Ghizzawi, the baker from Afghanistan, with a practiced rhythm, running through his medical condition, the CSRT reversals, and his prospects of surviving another year.

Peter Karafotas, thirty-one, deputy chief of staff for Schakowsky—a liberal Democrat from the Chicago suburbs and mentor to Obama—finally asks, "What can we do?" It's not dismissive or resigned. He really wants to know.

Candace says that Schakowsky, who is both her congresswoman and a member of the House Intelligence committee, should look into Ghizzawi's case in her official capacity. Her status on the committee allows her to see classified materials.

She starts loading down Karafotas and his even younger legislative aide with documents: the famous Abraham Declaration; another declaration, out last month, from an anonymous tribunal judge who sat on nearly fifty panels and arrived at similar conclusions; and her filings to the Supreme Court, which detail her client's medical woes and claim violations of basic protections.

"How, again, do you know the congresswoman?" Karafotas asks, and Candace tells a funny story of how Schakowsky helped sponsor a show that she organized with some other detainee lawyers in April 2006. The cast of a British production, *Guantanamo: Honor Bound to Defend Freedom*, was flown to the United States to perform their edgy play of lawyers and detainees in the Foyer of the Rayburn House Office Building. "At the start, I asked Jan, can we really do theater in Congress, and she said, 'Absolutely, sometimes it seems like all we do here.' All I can say is, we had a great turnout for the show."

The aides say they'll pass everything along to their boss, and she'll bring up Ghizzawi's case on the Intel committee. But Candace isn't hoping for much.

While everyone waits for the seminal habeas corpus case to be argued before the Supreme Court in December—with a ruling probably not coming until June of 2008—there are no fundamental rights for Schakowsky, or any lawmaker, to grab hold of. It's all just political theater.

Then Candace is out the door, her smile gone. She has other offices to call on in the House and Senate. "I feel like I'm going door to door, doing a one-woman show," she says archly. In her briefcase, giving it weight and giving her a sense of righteousness, are letters Ghizzawi gave her during her last visit, on September 25. He had, essentially, written up his will and put it in an envelope.

He looked worse than ever then, and he told her he didn't expect to live much longer. One note elucidated all that had been done to him—the deprivations and torture—and requested compensation for his family. The other was a note about the disposition of his personal effects and his hopes for his daughter and wife.

He showed Candace two pictures of his daughter—one at two months old, when he was still with his family, the other when she was just a little older, in a snowsuit. He said he'd received other photos—pictures taken during the years he's been gone—where she's older, mugging for the camera, or jaunty, hands on hips. They were of a girl he'd never seen, and whom, he was now convinced, he'd never meet. "It was too painful for me to look at," he said, "I had to tear them up."

Candace said if he died, she'd do her best to help his family. He thanked her, and she pulled some documents from her briefcase. They were copies of various Supreme Court filings related to the habeas case soon to be argued.

"The injustices done to you, in a way, made this all happen," she said. "You may end up helping a lot people and helping this country get back on course."

He smiled wanly. "That's very nice," he said, with mild interest, flipping through the documents.

He saw his name. ABDUL HAMID ABDUL SALAM AL-GHIZZAWI.

"That's not right. That's not my name."

Candace froze. Not his name? She'd filed dozens of legal documents to U.S. courts, all the way to the Supreme Court, with the wrong name?

"Um, well that *is* a problem," she said, wondering if she should give him a primer on basic legal standards, dating back to ancient common law: that those who appear before a judge must use their proper name.

"What, may I ask, is your name?" she said, feeling the chill wind of suspicion she thought had long passed and would never return. Is it possible, after all they'd been through, that she didn't know who he was? Was he really even a baker, or a man who dreamed of his daughter?

"Okay, let me show you," Ghizzawi said, spinning the document around—her document—so he could explain. "There, you see, my name is Abdul Hamid al-Ghizzawi. Those other two words—*Abdul Salam*—are stuck in the middle." Ghizzawi looked up. "That's the name of the man in the cell next to me in the beginning, when I first came here. He's gone. He may have died. I don't know. But they made us one person."

She was speechless.

That's why Candace Gorman—mom, lawyer, soft-edged middle-aged American—is walking these halls, barging into congressional offices to perform her one-woman show.

She has no choice. Because in September of 2007, she saw it, the kind of invisible bond, really, that suffering or injustice or power, in its dark, punitive form, can create.

What happened is that she, the pencil-thin baker from Afghanistan, and the vanished man in the next cell all became one person.

———

Usman Khosa pulls out of a deep sleep, his cell phone ringing. He wakes up and grabs it.

"Hello," he croaks.

"Usman, turn on the TV." It's a college friend, a kid from Pakistan who lives in New York. "Everything's going to hell back home."

Usman leaps out of bed and flips on CNN. The lead story for Saturday morning, November 3: Musharraf Imposes State of Emergency in Pakistan.

He picks up the phone and calls Lahore.

"Aba," Usman, says, using the colloquial for *father*. "It's me. Are you okay?"

"We're okay, Usman, but it seems like a lot is happening here," Tariq says, and lays out what he knows, which is not very much. "Musharraf seems to have tried to take over the judiciary. That's clear."

He tells Usman that he's at the home of his brother, Asif, a judge of the Lahore High Court. Tariq says that all the Khosas have gathered there, which makes this suburban home in Lahore a sort of bunker for the loyal opposition to dictatorial rule. The Khosas are an extended family at the core of Pakistan's growing professional class—their clan is filled with lawyers, judges, police officials, and longtime government ministers. Latif Khosa, Tariq's first cousin, was the legislator at the head of the lawyers' revolt, literally: the photo of him in a clash with police, defiant, his head bleeding, circled the globe as the movement's defining image. Across two generations since the country's founding, this family—and a few others like it—have fought to preserve Jinnah's progressive ideals about education and merit and rule of law, even under extreme circumstances. That has often meant making compromises to survive. Judge Asif Khosa's elderly father-in-law, who lives next door, is a former chief justice of the Pakistan Supreme Court. In an earlier era, under another military dictator, General Zia ul-Haq, he gave judicial approval to the execution of Zulfikar Ali Bhutto. Years hence, his relative by marriage, Latif Khosa, is Benazir Bhutto's lawyer in Pakistan.

And now, as another crisis of law versus power unfolds, Tariq tells Usman he should call back immediately on his uncle's landline. He does and talks to his aunt. Along with the state of emergency, there's a news blackout across Pakistan, and the Internet is blocked—most likely on purpose, by the government. Everyone is in the dark.

"Let me find out what's happening," Usman tells his aunt. "I'll call you back."

He slips on some gym shorts, opens his laptop, and starts trolling news websites and video links.

Two floors down, the door to the street opens. It's

Sadia, his girlfriend. They're supposed to go to brunch
this morning, do a little work, and then wander around
the city. Things have been getting more serious between them. Though Usman hasn't formally proposed,
it seems as though they're moving toward marriage.
Later this month, he's supposed to meet her family
in Illinois. And he's been talking a lot about big life
issues—his goals, his career, how he'd like a lot of children.

"Usman?" she calls from the stairs.

"Sadia? Come up. Hurry!"

A moment later, she's on his bed, her laptop, which
she carries everywhere, open. Usman trolls the BBC
online. Both of them are trying to hook into Geo TV, a
private Pakistani television station.

The situation slowly comes into focus. Two hours
before—at 7:00 a.m., Eastern Standard Time; 5:00
p.m., Saturday afternoon, in Pakistan—the television
stations across the country went dark. Government
troops appeared outside the offices of the Pakistani Supreme Court and the country's two major provincial
high courts—the one in Lahore, which covers Pakistan's economic and cultural center, Punjab Province,
and the court in Sindh Province, which oversees the
country's impoverished southern half, including Karachi. The judges were told they must sign an oath sup-

porting a "provisional constitutional order" allowing Musharraf to establish a state of emergency. If they refused to sign, they'd be taken to their homes and placed under house arrest.

Pakistan's judicial and executive branches have been heading toward a showdown since the late winter, when Musharraf's suspension of the Supreme Court chief, Iftikhar Chaudhry, sparked nationwide protests.

Bhutto's chaotic return—the lead-up, the terrible attacks, and the weeks of follow-up accusations about government involvement—then embroiled the country in a set of parallel controversies. The three major players—Musharraf, Bhutto, and a reinstated Chaudhry—created a complex push and shove. By not withdrawing her party from the legislature for its October 6 presidential vote, Bhutto lent credibility to the general's reelection—a managed parliamentary plebiscite—to another five-year term. That bought her passage for a return that should have been triumphal, but that instead left the streets bloodied and the country locked in vicious recriminations.

At that point, Chaudhry and the Supreme Court stepped up, saying they would soon issue a ruling on the legitimacy of the October 6 vote. By Monday, October 29, Musharraf was convinced their ruling would go against him. Facing a vote from the Court early in the coming week, he reviewed his options and weighed the idea of calling a state of emergency. Rice caught wind of the possibility and called him, saying the U.S. would frown on any extraconstitutional moves. But, once again, Musharraf knew the United States was bluffing.

By week's end, all forces were fast converging. A concerned Admiral William Fallon, the U.S. military chief for Central Asia and the Middle East, visited Musharraf on Friday, November 2, full of bark but without the bite to match it. If the United States really wanted to give Musharraf pause, Fallon would have been carrying a list of sanctions: specifics about the withdrawal of U.S. aid in the event that martial law were declared. He had nothing in his hand.

Musharraf, on November 3, finally moved. All his talk over nearly eight years about wanting to shepherd the country toward democracy was, in an instant, shown to be hollow.

Back in America, Usman and Sadia try searching for video links to Geo TV. It's shut down inside Pakistan, but it seems to be putting out some sort of international feed. They manage to connect a few times but it keeps shutting down. Usman tries another site—PakVideo.com—and connects.

Sadia is meanwhile picking through some reports from Indian journalists. Between the two of them, they have fourteen windows open on their computers. CNN is blaring in the background.

Ten minutes pass. Loaded up with disclosures, Usman calls Pakistan.

His uncle's house is now crowded with family friends, other judges, and officials of the government. Tariq picks up the phone.

"It's Usman, calling from America," he shouts. People crowd around the phone with questions: "Ask him, what's happening?"

Usman passes along the latest reports. Journalists have been arrested, one report says. The radio stations are also closed down.

Sadia, sitting next to Usman, shouts, "There's an Associated Press report that political opponents are being rounded up."

Usman passes this along.

His roommate Linas, now awake after a night of partying, is soon recruited. He plugs in his laptop and flips to the BBC on the living room TV. He yells up, "There are spontaneous demonstrations in the streets."

Usman passes it to Pakistan—"Aba, there are reports from the BBC of riots"—and then he calls down to Linas, "Do they say where the riots are occurring?"

This goes on through the American morning—the Pakistani evening—as disclosures and breaking news flow from Dupont Circle's outpost of the Pakistan Government in Exile to the Lahore safe house. The Khosa brothers then call friends and colleagues throughout Pakistan, relaying the news from their improvised Washington feed.

Soldiers surround the house in Lahore. The thirty-five top judges in Pakistan, including Chaudhry and Asif Khosa, are now officially under house arrest. Inside there is heated debate. Should Asif and other judges sign the oath, return to their offices, and then declare Musharraf's acts illegal? Does the constitutional order cover all laws? Usman calls in with more fresh reports; the Khosas call others. It's like a crack in a dike, water gushing through.

At noon, Usman picks up a Geo TV confirmation that four of the Supreme Court justices, known supporters of Musharraf, have taken an oath to uphold the state of emergency. Abdul Hamid Dogar, a Musharraf loyalist, will take Chaudhry's place as chief justice.

It means that a captive judiciary branch has been seated, and that the refusal to sign the oath will be the last act of most of the country's judges. Usman calls Pakistan. Tariq puts his brother on the phone.

"I'm sorry to be the one to tell you this, Uncle Asif," he says and repeats what he's just learned. Asif Khosa has been a high court judge—a source of enormous pride for the family—since 1998. He knows he may never sit on the bench again.

"It is better I hear it from you, Usman," he says. "I'm not budging. It's not worth changing your principles and signing that oath. But it's hard to hear this news."

Of course it's better that he know, so that he and the other Khosas are not kept in the dark, feeling powerless about the fate of their country.

Usman hangs up and turns to Sadia. He's made eight long calls to Pakistan since he was startled awake. "Sadia, do you have any more credit on your phone card? Mine's out." Hers is out, too, so she goes online, taps in her Visa number, and starts downloading cash to Usman's phone card. She does the same for herself, thinking that someday, maybe, they'll tell their kids about this—if marriage or kids or any of that happens. She flips open her cell phone and orders pizza. She figures they'll be there all day, and all night.

Digging up disclosures and trumpeting them to those who need to know was once a satisfaction known mostly to journalists. That sensation—felt by Usman and Sadia—is also felt, nowadays, by the airline fanatics who tracked public records of flight manifests and helped uncover the U.S. program of extraordinary rendition, the illegal transfer of American captives between countries and, often, to secret prisons. It's felt by the Egyptian bloggers who've been dogging the Mubarak regime, and after they were arrested, by the next wave who quickly replaced them. It's a sensation that guides the online hordes who raced to download the list of Guantánamo detainees in early 2005, after the names were made public by Lieutenant Commander Matthew Diaz, who was summarily court-martialed and imprisoned. The download community then fanned out, uncovering what it could about each detainee.

What you might not expect is that many of those in this transparency-minded community are not notably strident about particular principles, such as detainee rights. For them, it's about secrecy. It affronts people. It makes them feel excluded and underprivileged. Almost by reflex, they push against it.

Which is why the U.S. government has been pushing in the opposite direction with such force. Vice President Cheney's famous quote after 9/11—about the United States having to operate on "the dark side" to fight a new kind of war—comes out of a belief that awful things must be done when dealing with a bloodthirsty opponent and that officials should be trusted to do what's needed, without having to answer to Congress or to the public. He's a firm believer in this. Since

Bush arrived in office, Cheney and others have fought to push significant functions of the U.S. government under the table and keep them there. It's policy.

It's also the reason the administration first hid the expansion of NSA's wiretapping latitude after 9/11, and why it has worked relentlessly to keep the details of this program under wraps since its disclosure. Those widened powers are not, however, just for catching terrorists. A prime purpose is also to see what's in everyone's hands at geopolitical poker tables around the world.

Unlike Rice, Cheney knows—after hard lessons across thirty years—that the United States often loses it shirt in the back rooms of the Middle East, the Persian Gulf, and South Asia because it can't fathom what's happening under the table. Sometimes it's a matter of misreading regional issues or long-standing disputes; sometimes the regional players secretly ally against the interloper, and set traps to doom U.S. initiatives and pick America's wallet clean.

The initiative to reinsert Bhutto into Pakistan was, in fact, launched and led by Rice and her State Department.

Cheney's position, expressed to the president on several occasions, was "don't mess with this," according to one of his senior foreign policy advisers. "Our feeling," said Cheney's adviser, summing up the view of the vice president, "was that arranging this marriage can only backfire on us. Bhutto is complicated and unpredictable. It's best to just support Musharraf, give him whatever he wants or needs to stay in power."

"Our position," the adviser added, "is that this whole thing with Bhutto is being run out of State. Let them fly or fall on their own."

Benazir Bhutto knew several days in advance that the state of emergency was coming. She had good sources and heard on Monday that Musharraf consulted with members of his legal team about how to draw up the emergency order. On Thursday, Bhutto took a plane to Dubai, saying she'd wanted to go see her ailing mother and her children.

Mostly, the trip was to get her affairs in order for what she knew would be a long, tough struggle with Musharraf. It had become clear to her that she would now have to meet the general's show of dictatorial power by leading some sort of public challenge. Flowery speeches are one thing. Taking a stand on democratic principles is quite another, especially when it might spark, or even depend on, civil unrest. She'd always been more comfortable cutting deals in the back room than leading marches. Democracy is messy. It happens in sunlight. It can get you killed.

On the late afternoon of November 3, Bhutto makes phone calls to loved ones and close advisers to get her affairs in order.

The NSA is listening. They've been listening to her calls for months, including an earlier call she made to her son, Bilawal, a bright, bookish boy and a top student at Oxford, who shares his mother's distinctive features: her full lips and wide, soft eyes.

In that call, she told him about the secret bank accounts that hold the family's fortunes—huge reserves of money that investigators have long suspected are ill-gotten.

When a precious secret is collected—as the NSA has done here—it tends to glow in the darkness. Placed in daylight, fitted along a wide landscape of facts, it often loses its brilliance. But that rarely happens. Secrets spoil in open air, they lose their potency, so they are sealed away. The subject of the secret is often aware that evidence has been collected that may be used to drive judgments and maybe even destructive actions. Once the secret is locked in a safe, the subject, in essence, never gets to exercise the ancient right to face one's accuser.

On this day, November 3, two competing ideas about the power of information arrive at a head-on collision.

On one side, Musharraf is busy rounding up judges

and political opponents while trying to keep the citizens of his country in the dark. This control of information, he hopes, will allow him to exercise power without any semblance of accountability.

The NSA, meanwhile, has harvested a number of portentous conversations of Benazir Bhutto's. This should help the United States play its under-the-table, cutthroat games more effectively. The intercept will be cited inside the U.S. government as evidence of Bhutto's unfitness, her corruption. It will be used as part of a wider, "carrot and stick" program begun in the spring, in which the United States let Bhutto know they were happy to work with her in setting up a marriage with Musharraf, but that they could make her life difficult—in terms of her finances and lingering corruption investigations—if she started to improvise and freelance. What they'll overlook is the context and her tone in the many calls they eavesdrop on—overlook the fact that she's scared and preparing for the possibility of imminent death.

But today's trend line, thankfully, belongs to Usman and his friends, who are freely relaying to some of Pakistan's top officials everything they need to know, in defiance of the repressive, secrecy-obsessed urges of those in power.

If the blistering disclosures of the past few years

about once-secret activities of the United States prove anything, it's that the most forward-thinking policy is simply to refrain from doing anything that you would be ashamed to have revealed. Because it will be. And there will then be a sense of disappointment—betrayal, even—about the gap between word and deed, especially if the latter is of the cutthroat, backroom variety, meant from the start to be concealed. It's this gap that kills moral authority.

The fact is that secrecy oaths, like news blackouts, are just another form of command and control whose time is rapidly passing.

That's the message from the apartment in Dupont Circle that became Pakistan's thoroughly unofficial Government in Exile. What they know, like young adults far and wide, is that the real future is open source—bathed in the light of day. It's the way the world is moving. Eating pizza late on November 3, this small global communications team will regale each other with the ins and outs of their amazing day.

And by day's end, Sadia is almost certain that in years to come she and Usman will tell their kids all about it.

Rolf Mowatt-Larssen didn't leave the government after the big August meeting in the Situation Room. He thought about it, and he's thinking about it now, driving up the East Coast from Washington to Boston a few days after Thanksgiving. He has to give a speech tomorrow at Harvard. Lots of important people will be there—noted professors, former government heavyweights. He's not sure what he's going to say, not sure how candid he should be about what he really knows and feels.

There have been a few months of twists, turns, and uneven progress since Bush was officially alerted as to how little has actually been done in the battle to keep uranium out of the hands of terrorists. But no action plan has emerged—nothing close—and Rolf's plan B on his "Armageddon Test" is looking shaky. His silent partner, Rob Richer, is going to be otherwise engaged for a while. Blackwater, which Rob and his intelligence firm are still part of, has become embroiled in a searing controversy since mid-September, when its operatives allegedly gunned down seventeen Iraqi civilians in Baghdad's Nisour Square. Investigations and congressional inquiries are already under way. Erik Prince, Blackwater's founder, has become a household name, and the Iraqis are calling for prosecutions and reparations. Richer and Blackwater's vice chairman, Cofer Black, the former CIA counterterrorism chief, were called in to do damage control by staying in close touch with former colleagues inside the administration. A great deal was at stake financially. Blackwater had collected more than $800 million in fees from the State Department alone for security contracting work, mostly in Iraq.

As for the plan to launch teams from within the government, Rolf is still looking to collect on his bureaucratic deal. He agreed not to criticize CIA, or its operational deficits, in the meeting with Bush—a precondition for letting him represent the entire intelligence community, on the issue of nuclear terrorism—even if it meant the Bush meeting would be more about generalities than hard specifics and finish with an open-ended "we'll get back to you, Mr. President." The second, unspoken, part of that deal was that Rolf would help get that so-called deliverable for Bush by creating "Armageddon Test" teams with CIA.

As September passed into October, it seemed CIA wasn't holding up its end of the bargain. No one wanted to meet with him. On October 2, Rolf's frustration boiled over at a Homeland Security conference at the Broadmoor Hotel in Colorado Springs. After he gave his latest alarming presentation about the nature of the threat, the floor was opened for questions. Rolf got one from a ringer: a freshly minted contracting executive named John Gannon, a twenty-four-year CIA veteran, former deputy director for intelligence, and once head of the National Intelligence Council, the group that puts out National Intelligence Estimates.

"Okay, Rolf," Gannon said, you're telling us all this harrowing stuff we now know. And this is a big deal, but the question is, "what are we doing and are we doing enough?"

There's an official response to that sort question, along the lines of: "We're doing a lot, but we could do more, and here's what we have planned." Anything short of that, especially on an issue Bush has called the "number one priority," is news. Rolf started to do the usual waffle, but then he stopped. He couldn't manage it.

"No, we're definitely not doing enough," he told Gannon. Rolf was thankful that there were no reporters in the room.

A few days later, all the top officials of the intelligence community gathered with members of the Senate's Select Committee on Intelligence for a two-day retreat in Virginia. Steve Kappes, the deputy director of CIA, addressed the group on the first day. Rolf huddled with him after the speech.

"You've upheld the honor of the intelligence community," Kappes said, archly. "The price we're going to pay for that is that we're actually going to have to do this uranium thing."

In the coming weeks, there were a few meetings with CIA clandestine chiefs, but Rolf sensed resistance of the traditional bureaucratic variety. This is our turf: we run the clandestine work at CIA, and we're not going to be directed in an operation coming out of Energy. Fine. The next day, Rolf got clearance from Sam Bodman that put him and Jim "Mad Dog" Lawler, the man who'd caught A. Q. Khan, on loan to CIA in their spare time, off ledger, just to help. It would be CIA's operation, for better or for worse. Rolf and Jim would be in an oversight role, structuring the operation, keeping it on course.

That's where things stood from late October until now. There was nothing else left to give. Rolf had upheld "the honor of the IC" with Bush, and now was offering his services for free. And CIA was still not responsive. They were discussing some sting operations. The U.S. intelligence community has been engaged in stings on uranium purchasing since the mid-1990s, with very little success. But, no. In terms of an audacious effort to send out teams to infiltrate the world's uranium markets—to get the uranium off the market, and identify both buyers and sellers—nothing.

Which is what Rolf is thinking about as he drives his Ford Taurus up I-95, bound for Boston. He could have done the usual: step on a plane, give his speech, and be home for dinner. But this is different. He felt he needed to get away from the crush of bureaucracy, the endless meetings, the tactical struggles, and clear his head. So he rented a car and brought the family. They'd drive up and, in a few days, fly home. Rosie, his wife, is with him, and his daughter, Kelsey, a brainy blond high school student who's going to attend a Harvard Law School class with a professor friend of the family.

Interstate 95 going north is jammed on this Sunday night at the end of Thanksgiving weekend. Rolf doesn't mind the traffic. He's moving slowly but steadily, and the long drive gives him a chance to think.

"I'm more depressed about the realities of the mature bureaucracy and more excited about the ideal that we're fighting for," he says. "And that juxtaposition has obvious consequences."

The decision to leave public service sometime soon, after thirty-six years, has freed him. This happens so often with long-standing government officials, who at the end of their time in public service can finally look at the surrounding terrain with new eyes—those of someone about to leave. "Americans are incapable of accepting that everything doesn't have a cause and effect, it's just the way we are," he says. "Therefore, 9/11 is the failure of something—it has to be—something that can be identified as wrong and be corrected. That's exactly what our opponent is counting on. They'll move in a completely new way the next time, there's no doubt." He pauses. "Take the Muslim world. What are they expecting of us? Not to understand them, maybe. To treat them all as one thing, one monolith. To not view them as peers—lots of them—because we're wealthy and smug and always right. Exactly. So we continue to confirm their expectations. But how could we move in another direction, to surprise those expectations, confound them, to constantly be moving on a fresh landscape, like our opponents? What's the first thing you'd do? And the second?"

He knows he's slipped into a dangerous state of mind, but it's envigorating, revitalizing. He doesn't know what he'll say tomorrow at Harvard. Or at another speech he has scheduled a few days later at West Point. Or another one next week, to the large anti-terrorism division of the New York City Police Department, which runs it owns operations—often flouting the FBI and federal government—like some small nation-state. One thing's for certain: he'll be way off message.

His wife and daughter are asleep as the road begins to clear just north of New York—how many times, he wonders, has he run models showing the destruction of this city?

Then he thinks of Ezekiel.

This has happened before, but it's been quite a while. There's a passage from the book of Ezekiel that's been rattling around inside him since a bizarre moment in the 1970s, when he was out drinking with some buddies from West Point and a man approached their table. He looked right at Rolf and said, "Read Ezekiel!" Rolf was raised a traditional Lutheran, though he wasn't much into religion as a boy or a young man, but the moment was so strange he went ahead and found a Bible. It was Ezekiel 33:1–6 that grabbed him, and stuck.

In it, the angry Old Testament prophet warns, in a voice he claims is from God, that the people should select someone as their "watchman," who will see "the sword coming against the land" and "blow the trumpet to warn the people." If he does, then, in essence, the people are responsible for the own actions, because they have been warned. That's not the part that gnaws at Rolf. It's the last part, where Ezekiel says that "if the watchman sees the sword coming and does not blow the trumpet to warn the people and the sword comes and takes the life of one of them . . . I will hold the watchman accountable for his blood."

In the years since 9/11, Rolf, like so many people in the intelligence community, has struggled mightily to simply see "the sword coming," to detect the imminent threat—to counter it—while the job to "warn the people" has been left to their bosses, the duly elected leaders. The lack of trust between these two groups, however, has grown, year by year, with the duly elected feeling they've been undermined by the intelligence community—by officials who've repeatedly claimed that the people have been misled.

On this cool, starlit evening, Rolf Mowatt-Larssen starts to see how the equation has flipped. There's no way, at present, to see "the sword coming," even in this most pressing area where terrorists and nuclear devices intersect. If the unimaginable were to happen, God forbid, the country would surely slip into a defensive, vengeful posture that would undoubtedly deepen global strife in precisely the way Islamic radicals have long dreamed about. Holy war with nukes.

As the highway passes beneath his wheels, and the East Coast of America drifts off to sleep, Rolf sees it: a future. His job will be to "warn the people" that we don't know what we need to know, that unless we reestablish America's moral leadership—to generate precious intelligence and global action—we will never see that sword coming.

"**Mr. Chief** Justice, and may it please the court."

Seth Waxman's voice echoes through the Supreme Court chambers, across the hushed audience, the row of justices, and the heavy neoclassical columns behind them.

"The Petitioners in these cases," he says, "all have been confined at Guantánamo for almost six years, yet not one has ever had meaningful notice of the factual grounds of detention or a fair opportunity to dispute those grounds before a neutral decision-maker."

With that, the head lawyer for the consolidated cases, *Boumedeine v. Bush* and *Al Odah v. United States*, begins oral argument and the third major test of the administration's policies on detaining what it considers "enemy combatants" in the prison camp at Guantánamo Bay.

The last sentence, about hearing the "factual grounds of detention" and being given a "fair opportunity to dispute" them, is, in essence, Waxman's citing the writ of habeas corpus—or at least the rights he will argue the foundational writ guarantees.

The term "habeas corpus," a Latin command "that you produce the body"—meaning that you bring the individual in custody before the court—refers to the writ, or legal procedure, by which a prisoner can challenge the legality of his detention. The writ's origins are found in early common law and date back at least to the Magna Carta, if not further. Though it is a principle at the very foundation of all modern legal systems, it is mentioned only once in the U.S. Constitution. Article I, Section 9, states that, "The Privilege of the Writ of Habeas Corpus shall not be suspended, unless when in Cases of Rebellion or Invasion, the public Safety may require it."

The Court has already looked at the question of these detainees' habeas rights in several cases. Each case turned on the extent of the administration and Congress's license in suspending the writ for Guantánamo detainees. The Court twice said its sister branches had taken too much license, and twice the administration went to Congress to produce statues reaffirming its detention policy. That resultant legislation—the Detainee Treatment Act of 2005 and the Military Commissions Act of 2006—explicitly took habeas off the table, saying that the administration's current procedures, relying on the CSRT panels and annual reviews, constituted an acceptable replacement for habeas rights and prohibited federal courts from hearing habeas petitions coming out of Guantánamo.

The U.S. appeals court in D.C., known as the D.C. Circuit, upheld this provision in the ensuing year, and the Supreme Court refused to review its ruling. That was until Stephen Abraham's declaration took a hatchet to the government's claim that the military tribunals provided an adequate substitute for the Great Writ, habeas.

The larger question at play in today's case is about whether Congress can deny the courts the power to hear habeas claims. That makes this, even more than the preceding detainee cases, a fundamental constitutional clash between branches of the U.S. government.

That's why, on this rainy day, December 5, the oral arguments are being presented to a sellout house. Many of those in the hallowed chamber lined up since the previous night to get one of about a few hundred precious seats in the audience. In the front rows is the entire Washington establishment, from noted judges to several dozen senators, congressmen, and assorted luminaries, all of whom had their assistants call for seats well in advance. In this case, some even called themselves.

A few sentences into Waxman's opening, the justices start to interrupt with questions, and it's clear they can't agree on a starting point.

Chief Justice John Roberts leads, asking whether the Court hadn't previously ruled, in *Hamdi v. Rumsfeld*, that the procedures laid out in the Detainee Treatment Act were adequate for American citizens, much less—it is implied—detained aliens.

Waxman counters that they might have been adequate had they incorporated certain basic legal rights—what he calls "process minimums"—such as "meaningful notice of the factual grounds for detention" and "a meaningful opportunity to present evidence in response to that before a neutral tribunal with the assistance of counsel."

Waxman is firm: "That's not what they got."

Ruth Bader Ginsberg chimes in next, wondering whether *adequacy* is even the issue before the justices today. The lower courts, relying on the congressional legislation striking down the prisoners' habeas rights, hadn't even looked into the adequacy of the habeas replacement. "So, shouldn't we . . . send it back to them to make that determination: whether habeas being required, this is an adequate substitute?"

Never to be outdone, Antonin Scalia takes Ginsberg's pragmatic line of questioning to a more fundamental one. Scalia is considered a leader in the camp of strict constructionists—judges who cleave to the language and original intent of laws to fight against what they consider inappropriate judicial interpretation or "activism." They criticize "judges who legislate," a role they say belongs to our elected leaders, even if the latter may sometimes act in ill-considered ways. His kindred on this court—Chief Justice Roberts, and Justices Samuel

Alito and Clarence Thomas—find their philosophical opposition in Justice Ginsberg, along with Justices John Paul Stevens, Stephen Breyer, and David Souter, who see this judicial conservatism as an at times problematic constraint on the fundamental power of the law to evolve and thereby meet the needs of successive eras in American life. In the middle of these two groups is Justice Anthony Kennedy, a free agent who often casts the deciding vote in major 5–4 decisions. For this reason, some joke that this is the era of "the Kennedy Court."

With Waxman up, arguing that the United States should provide habeas rights to the diverse assortment of foreigners housed at Guantánamo, Scalia can barely constrain himself.

"Do you have a single case, in the 220 years of our country," he says, his voice rising, "or for that matter, in the five centuries of the English empire, in which habeas was granted to an alien in a territory that was not under the sovereign control of either the United States or England?"

Scalia, of course, is matching a strict constructionist, precedent-driven argument to the exceptionalism of Guantánamo, chosen by administration lawyers precisely to be outside of precedent. Aside from those in active war zones, it's the only American base beyond the sovereign United States that is not subject to a "status-of-forces agreement," the common legal framework that mediates between the U.S. law and that of the host country on bases abroad.

Guantánamo Bay, one of a kind, was seen as the ideal detention center for a new kind of war. What the administration was counting on in choosing it—that no clear laws would apply there—is what Scalia now shapes into a legal argument.

He and Waxman rhetorically chase each other around the courtroom, as various barely applicable precedents from Britain's colonial period—strangely named cases like *Earle of Crewe*, the *Case of Three Spanish Sailors*, and something called *Mwenya*—are debated.

As the argument unfolds, however, a hidden weakness on the issue of precedent is shown in the very uniqueness of Guantánamo: if there's no precise precedent to show that habeas *should* be extended to foreign nationals at Guantánamo, there's none to show that it *shouldn't*.

Waxman turns the unrestricted authority at Guantánamo Bay—the very reason it was chosen—against its defenders. The "complete, utterly exclusive, jurisdiction and control of the national government of the United States" on the base, he argues, makes it uniquely subject to U.S. legal obligations.

"Our national control over Guantánamo is greater

than it is over a place like Kentucky," Waxman explains, which has state laws that compete with federal statutes. It is precisely the administration's selection of a place where it can have absolute, unfettered authority over the detainees that erodes its contention that such a place is beyond U.S. law. The idea is that American authority carries with it American law.

An unspoken current running through the proceedings concerns the Framers' *intended* fetters on central authority and their relevance in a place the Founding Fathers would have had trouble imagining. Of course, this isn't just about the fate of three hundred Muslim men caught in the crosscurrents of a new and messy struggle. It's also about executive power and the ability of the courts to push back against the other branches of government when they overreach. Habeas, from the very start, has been an essential check of law over leader.

As the oral arguments proceed, it moves past the issue of whether habeas applies in the matter of Guantánamo. Taking the Great Writ as a given—and asserting the role of the Court in defining and defending habeas—the justices begin to look into the adequacy of its replacements.

Kennedy finally weighs in. When he speaks, everyone, including the chief justice, listens.

"Suppose there had not been a six-year wait, would it have been appropriate then for us . . . to defer until the court of appeals for the District of Columbia" finished reviewing the various cases from Guantánamo on its docket? The D.C. Circuit can review CSRT findings and make certain circumscribed determinations about whether they comport with U.S. laws, but Waxman says the process has been stalled for two years, with no immediate prospect for moving forward.

Kennedy is digging toward the core of the current appeals process, to find out why it's not working. After several exchanges, the line of questioning arrives at the central issue of evidence and the inability of any appeals court to review CSRT decisions in a meaningful way without those "process minimums" that Waxman raised with Roberts in the day's first exchange— particularly a detainee's inability to present evidence and "the presumption" that all the evidence provided by the government is "accurate and genuine."

Fine, says Kennedy. Why can't this problem, these issues, simply be taken up in "the CSRT review proceedings that are pending?"

"It could," Waxman responds, "if the military had different procedures to govern the CSRTs."

This is, actually, a key moment. Kennedy, the Court's swing vote, is asking, essentially, can the current system be fixed to comport with habeas. Waxman says no. After years of tinkering, the Guantánamo system is broken, an ensnaring maze. "I would say with respect to future detainees, that this Court could issue a ruling—well, this Court should issue a ruling—saying for these people, if the writ means anything, the time of experimentation is over. We have tried and true, established procedures," he says, in the public courtrooms of the United States.

Kennedy takes this in. He has no further questions. Every justice—mindful of how Kennedy's mind often becomes the will of the Court—makes a mental note.

Waxman is followed by Paul D. Clement, the U.S. solicitor general, who tries to convince the Court that the alternative proceedings, such as the CSRT, are more than adequate to the spirit of habeas.

If they failed to provide meaningful legal review for the detainees, Clement says, the procedures are not necessarily to blame, but how the process was carried out. "The interesting questions that I think are left," he adds, "are issues about whether or not, based on the Abraham Declaration, that the military followed their own procedures."

Sitting fifty feet from Clement, in a small gallery of seats just off to the right of the row of justices, Stephen Abraham can't help but smile. He took a few days off from the pomegranate wars and other civil actions to fly here from California. In his best blue suit, he watches the country's top Court follow through on the process that he and his declaration jump-started back in June. He had to be here, to see how it all turned out. Abraham, of course, once believed—or wanted to believe—the very thing Clement is now arguing, that the problem was how "the process was carried out." Now he feels the entire thing was a setup, built from the start to stand in the way of justice.

The rights accorded under U.S. law, after all, are neither generous nor flimsy nor "more than adequate." They evolved, in a rigorous evolution over centuries, to be simply sufficient. What Abraham knows, from hard experience, is how removing one of these rights—such as the right to see the assembled evidence or the right to counsel—can throw everything into chaos, where seemingly sound procedures no longer make sense, and a proud legal system is thwarted, sent chasing its own tail.

At final rebuttal, after nearly an hour and a half of thrust and parry over legal procedures and their applications, Waxman—solicitor general under Clinton and one of country's most skilled legal tacticians—tries to summon absurdity. He realizes he needs to show how, outside the abstract realm of legal precedent and theory, the concrete system, currently adjudicating the fates of actual human beings, fails and has failed to provide basic, commonsense justice.

He tells, as quickly as he can manage, the story of Mr. Kurnaz, a permanent resident of Germany, who was told at his CSRT hearing the name of a man he was accused of having associated with—the basis of his detention. The man was, the government asserted, a known terrorist named Selcook Bilgen, who had blown himself up in a suicide bombing.

"All that Mr. Kurnaz could say at his CSRT," Waxman recounts, "was 'I never had any reason to suspect he was a terrorist.'"

When Kurnaz's lawyer filed his client's claim for habeas corpus, he saw the transcript of the CSRT hearing filed by the government, on which Bilgen's name appeared.

"Within twenty-four hours," Waxman says, all but spitting out the words as his time runs out, Kurnaz's lawyer had an affidavit "from the supposedly deceased Mr. Bilgen, who is a resident of Dresden, never involved in terrorism, and fully getting on with his life."

"That evidence," clearing Kurnaz, "would not have been allowed" in the legal reviews available to the detainees, Waxman concludes. "And that's why it is inadequate."

What Waxman did, at the very last minute, was make the robed justices all feel what the man sitting just to their right, in his best blue suit, felt when he looked into the case of Abdul Hamid al-Ghizzawi.

Outrage.

Chapter 3
Waiting for the Call

Usman Khosa is the second most indispensable man in Lahore.

The first is the guy marrying his sister.

In his role as the bride's able-bodied older brother—with a fast car, quick reflexes, and an aptitude for following instructions—Usman has to do more than just show up. A lot more. It's all there on a long list from his mother, starting at the top: "flowers." By sundown, if everything works out, he'll be hauling the last of his charges—desserts—in the backseat.

Tonight, December 12, is the Khosa Wedding. Everything must—everything *will*—be perfect.

Usman's younger sister, Sana, is preparing to marry a nice young man, Zain, the child of a family from Lahore, distantly related to the Khosas. As dusk arrives, the bride and groom will sit with an Imam, who will take them through a complex legal and religious dance. Then everyone will eat and celebrate into the sultry Punjabi night.

Usman arrived in Pakistan last night, after a harried exit from Washington and a twelve-hour flight. At a quick glance, it seems like he's leaping between worlds. He feels a bit of that today—a disconnected, displaced sensation—as he races around Lahore. This is not an issue of geography. He *is* of two worlds wherever he goes, and this is why he was so anxious to return to Lahore, to Pakistan, to work through some of the conflicts churning within him. It's something he feels he must do, as a young man fast maturing, and a Muslim. He's decided to marry Sadia. But first he needs to arrive at some reasonable clarity about where he, Usman Khosa, belongs.

Two weeks ago he told Matt McGrath, head of Barnes Richardson, that he was going back to Pakistan for a month, at least, and there was a chance he wouldn't return. Matt was surprised. He told Usman his job would remain open, and Usman said that he'd keep Matt posted.

"Promise?" Matt asked.

Usman nodded, "I'll let you know how it all turns out."

In his book *The Soul of Black Folks*, W. E. B. DuBois

writes about always feeling "his twoness,—an American, a Negro; two souls, two thoughts, two unrecon-
ciled strivings; two warring ideals in one dark body."
This sense of twoness, of course, has a wider global
signature. Identities based on race, ethnicity, nationality, class, ideology, and religion have always com-
peted against one another and—all together—against a
shared human identity. People have both searched for
self-definition and had it imposed on them, as flowing
migrations have, over time, multiplied the dualities—
the "twoness" and "warring ideals"—in every human
actor. The struggle over how individuals are defined
within nations and how nations define themselves is a
very long story, and a bloody one.

If a feature, in recent times, of America and its Western forebears has been to try to protect the process of
self-definition within their borders, providing individuals with the space to manage their issues of blended
identity on their own, then the extremist response, rising with such strength of late from radical Islam and
its global mission, is to force people to choose. Duality
is nothingness, they say. It's indecision—a confused,
often sinful search and an affront to their clarity, their
purity. Pick an identity; pick a side. Are you with us or
against us?

Of course, knitted into this challenge—one the

United States has issued to the world in what it calls the "global war on terror"—is the question of a greater *us*. Insofar as people are not all that different in their basic needs and joys, their yearnings and comforts, is there any force fighting for a shared human identity?

There are few better places to watch this unfolding conflict—between the forces of division and unity—than at the Khosa house in Lahore tonight.

That's not just because so many rivers of religion, nationalism, and ideology famously converge in Pakistan, but also because it's a wedding. The shared human identity, after all, is carried most reliably and most often through common experiences and rituals of family. Among all human ceremonies, there is nothing to quite match this celebration of choice—of a mate, a union, and a hoped-for future.

It's all here as the guests arrive, and Usman—having just unloaded from the car the *ghulab jamun*, flour balls dripping with sugar syrup—slips into his dark suit.

The house of Tariq and Ayesha Khosa is beautifully decorated for the day, an ivory tent draped across a wide courtyard. The house is large, like an upper-middle-class home in America, with five bedrooms and a high wall around its lawns and gardens. It's one of the perks of public service in Pakistan—a holdover from the Jinnah era and its reverence for civil service exam champions—that this house, owned by the state, is granted Tariq as he rises up through the ranks of the country's law enforcement bureaucracy. In fact, the neighborhood—filled with such houses and surrounded by its own high wall—is called "police town," a sequestered community where the country's law enforcement officials live side by side.

The families of all the Khosa brothers—Asif, the judge; Nasir, a top minister to Pakistan's prime minister; Arif, head of the trade association—and one sister, who runs a Lahore elementary school, are gathered at the eldest brother's house. An Imam, a friend of the groom's family, arrives, and the men greet him in the living room. There's an elaborate ceremony, where the groom signs documents and is asked three times if the intended marriage is acceptable to him. He says, three times, "acceptable," and is led by the Imam to a room where the bride awaits. She completes a similar ritual, signing documents as she's asked, three times, whether she is accepting the marriage of her own free will. With her three yeses, the two are married legally and in the eyes of Islam, and they sit side by side. All of this goes off smoothly. Usman's wisecracks at the start—a stage whisper to cousins that the Imam is dressed like Ayatollah Khomeini—give way to a quiet respect for the procession of ritual and emotion. Neither family is particularly religious. They're what would pass for the traditional mainstream, like New England Episcopalians, and soon the couple emerges into the courtyard, where about seventy-five family members and close friends gather to eat.

Usman makes his way to the buffet line. Crossing his path is a woman in all black. She'll take a plate and eat in private, where she can remove the veil that covers everything but her eyes.

This is Usman's beloved older sister, Sadia. She is in the full *abaya*, black gloves included. The veil shades her eyes. He looks over at her. "She certainly makes an entrance," he says, with a note of staged exasperation. "I don't see that much of her. Now we'll have a few days, at least."

Usman's mother, headmistress of a local school, is sweeping in and out, from the courtyard to the house and back, looking like a Bollywood actress surveying her all-star ensemble. Cousins have come in from far-off locales—this one at the University of Wisconsin, another at Cambridge, and several young lawyers from London. But, every few glances, her eyes check on Usman and Sadia, her two children who've returned tonight from across the world.

"One moved to London and became a fundamentalist. The other moved to New London and became a capitalist," she says, a favorite line, oft repeated in this group, which intently follows the fortunes of Sadia and Usman, sister and brother, so similar, so different.

Usman, in fact, desperately misses Sadia and the relationship they had for most of his life. He feels that she's vanished under her veil. They talked a few weeks ago, on the phone. It was a tense exchange. She wanted Usman to fly to London, pick her up, and then escort her to Lahore. Her husband, Imran, a young doctor, was planning to follow a few days later.

"Usman, it's not a question. Imran can't come, and I can't fly alone," Sadia had said, reminding him of fundamentalist Islam's injunction against women traveling long distances unescorted by a husband or male blood relation.

"This is crazy, Sadia," Usman replied. "You used to fly all over the world by yourself. This doesn't make any sense."

There was no resolution. For Usman, possibly leaving Washington for good, it would be battle enough to get done all he had to do and make it to Lahore in time for the wedding. Ultimately, Imran changed his work schedule.

Dressed in a formal black *kameez*, with a high collar, Imran now trolls the buffet table. Nearby, Asif, the ousted judge, is holding forth to brothers and cousins.

"I am collateral damage of the United States," he says, about his week of house arrest and the loss of his job. "I'm the direct result of them not supporting democratic institutions, like the judiciary, and instead supporting a dictator."

Then he describes what was happening behind the scenes during the two months before the state of emergency was declared, how the real struggle had been over the way "the intelligence agencies would pick up people, say they were terrorists, and then make them disappear. Chief Justice Chaudhry took up the issue. The Supreme Court simply asked the government lawyers, 'What are the allegations against many of these people?' They couldn't even say what the charges were, much less offer evidence. This is why things backed up at the courts in the last two months. Musharraf said we were releasing terrorists. It was ridiculous. We are a court. We live by laws. We lose that, we lose our soul. No doubt, when the ruling on his legitimacy was coming up, Musharraf sensed we'd go against him—that was a trigger. There was much more than that going on. It was a fundamental break on issues of rule of law. And how, people ask, can the United States support this man?"

He pauses, mobs his brow. "Even if you don't believe in the concept, why is it so hard for people to see that this bending of laws is a gift to the radicals?"

Imran, standing nearby, listens intently. Along with his medical work, he is a coordinator for Muslim student organizations. He lives inside the conflicts he now hears Asif, his uncle by marriage, discussing—conflicts over what constitutes suspicion, and whether suspicions should prompt action. He's on the receiving end. "So many of us are under suspicion in the UK for simply being the way we are: religious," he says. "Look, I'm the least violent person you could imagine, as are the students I advise. But we're targets."

To highlight this, he mentions an Islamic scholar who visited the UK from the United States a few years ago and was a mentor to Imran and his brother Asim—who is himself a lawyer and vocal Islamic activist who fights on behalf of the prisoners at Guantánamo and elsewhere. "It was a year after 9/11 and he was giving us advice about how to build bridges to mainstream Muslims and remove some of the conflict from the current religious debate." The scholar, a cancer researcher named Dr. Ali al-Timimi, would soon be convicted of inciting violence against the United States. His prosecution came out of a September 16, 2001, dinner he hosted, where he advised several young Muslim men from Northern Virginia to consider leaving the country and serving their faith as mujahideen. This was just days after 9/11, when Muslims across America feared for their safety. No U.S. troops were present or assigned to Afghanistan, and none of the men Timimi advised ever fought abroad. Nevertheless, in a case that spurred a bare minimum of controversy, the American-born Timimi was given a life sentence.

"A life sentence for speech," says Imran. "What could be a clearer violation of basic ideals?"

Standing next to him, Usman quietly sips a Diet Coke. Sadia lives in a house in London with Imran, his twin—a Muslim scholar who performed their wedding—and his brother Asim. Usman has met the brothers—likes them all—but is sometimes concerned about his sister living in a place of such fierce convictions. One of the brothers is a star on YouTube, where he appears in a video exhorting an Islamic rally in London to support jihadist efforts against "the oppression of the West."

"We all fall short of our ideals sometimes," Usman says to Imran. "The question is, do we learn from mistakes, and try to make amends?"

With that, Usman heads into the house, where the older men are gathered in the living room eating desserts from small plates. Here his father is holding court, talking about his children's successes in America—Usman's burgeoning career and the matriculation of Zain, Sana's new husband, next month at the University of Central Florida, where he'll pursue a Ph.D. in computer science. Sana, meanwhile, is about to start a master's program in public administration. "They are my harvest," Tariq says.

The reference is meant for the man nodding off in the chair next to him—Tariq's father, Faiz, who, a few months before, published a small book, a memoir, called *Harvest of Hope*, which he dedicated to "my wife and our five children and fifteen grand children who are our harvest of hope for the future of our great Nation." The book, a 116-page elegy, closes with a plea that "Pakistan will Inshallah emerge as a strong, vibrant, democratic and dignified country to fulfill the vision" of Jinnah. Tariq, who helped with the book's completion, has made sure all in this room have copies for themselves, their kids, and their friends. It's an offering he makes as an act of faith, a counterweight to a dispiriting turn in his life—and the life of his country as it whipsaws between extremist militancy and dictatorial crackdowns—over the past few months.

The talk in the room turns to politics and the upcoming election. There is much to discuss, of course. While preserving martial law through November and keeping political opponents, including Bhutto, under various states of house arrest, Musharraf has set a nationwide election for January 8. Critics charged that a fair election could not be held under a state of emergency, when Musharraf could simply and with impunity use government officials to fix it. Musharraf dismissed such concerns in public but, meanwhile, held planning sessions with a wide array of officials throughout November, including the inspectors general, who help oversee law enforcement in each of Pakistan's four provinces. Tariq, after several years as the number two official in Pakistan's version of the FBI, took over in January 2007 as inspector general of Balochistan, Pakistan's large southwestern frontier region, which abuts Afghanistan and Iran. It is home—particularly in its capital city, Quetta—to an increasingly radicalized Muslim population and a Taliban council that, remotely, manages and runs a significant number of the group's Afghan operations.

Musharraf's team told Tariq, in essence, that he was not cut out for what was ahead in terms of "electoral duties"—which involved using law enforcement officials help fix the election—and asked him to step down; the government would eventually find a role for him elsewhere. Though Tariq felt a kind of pride in being viewed as incorruptible, now, a month later, he's feeling a growing unease—in abeyance until when? Until after the election? Or until Musharraf locates some other job for him, if in fact Musharraf survives?

Of course none of these particulars, known to most of those assembled in the room, is discussed. The discussion is of radicals in the tribal areas, Musharraf, Bhutto, and America's part in everything—always that.

Then Tariq decides, after a few moments, that the country's politics are no subject for a wedding day. "Today," he says, "let's put all that aside for once." He wakes up his father—"Come on, Aba"—and motions to Usman, who's leaning in the doorway. Each man takes an arm and, together, they the escort the old man to a waiting car.

It's a sign for others to leave. When Usman and Tariq return, just the core of innermost family is gathered in the den.

A CD player emerges. The women sit on one side of the room. Sana is resplendent, arrayed in pale yellow. Her mother is in blue. Cousins, women and girls, sit around them, in chairs and on the carpet. The women, one by one, in twos and threes, rise to dance, and then recede as the men step forward. There is an elegance to this, the separations creating a back-and-forth rhythm, a tension of opposites that frames and accentuates the day's celebration of union, the coming together of a man and a woman. Soon everyone is part of it. The Khosa brothers, their lives thrown into disarray, move, side by side, with eyes shut and wearing soft smiles, their arms up, drifting forward and then giving way to a line of their concerned wives, now up and moving, all in a row, their traditional gowns and scarves lifting and falling to the music. They, too, are replaced, as the floor fills with the ambitious children they created, who've returned briefly from lives of storied triumphs—well burnished and at least half true—in England, the old empire, and in America, the new. Someone mentions Sadia, what a fine dancer she used to be, and so beautiful, and Mrs. Khosa nods; life changes, they grow up. Now Sadia is a pious wife, veiled in black, sequestered in an upstairs bedroom. Soon Sana will be going to America with her husband, following the long-lost Usman. But now they are here, all of them, with the music lifting and swelling. Tonight they dance in a celebration of change, really, a celebration that things change—for a family, a nation, all things—but the world still holds its moments of perfection.

The next day, Usman is traveling around Lahore, seeing his friends who've stayed in the country—some professionals, some running businesses, some still in graduate schools—and trying to get a sense of their lives.

In certain ways, their lives are not so different from his, not as much as he expected. The upper rungs in countries such as Pakistan—and this tier is growing, if slowly—are starting to look more and more American, or European, or Japanese, as they're drawn into the currents of globalized commerce and tastes. Usman has lunch with his first cousin Emmett, an entrepreneur who went to the University of Wisconsin, and his wife, who works as a consultant to the World Wildlife Fund.

As he makes house calls, Usman passes through the poverty and destitution that define most Pakistani lives. The streets in Lahore, Pakistan's most affluent city, are muddy from morning rain. There are beggars at every corner. Is there enough energy here, Usman wonders, enough forward motion? Some get rich, including a few of his friends, but life in the wide bottom remains unchanged. Can you call this progress?

In the afternoon, he drives by his sister Sadia's private girls' school and thinks of how she used to be the top student in Lahore, of how proud everyone was of her and all the items in the local papers. How many times had he heard his father tell the story of when Benazir Bhutto was sworn in as prime minister in 1988, and Sadia came down that morning for school dressed in Bhutto's green and white. "One day," she announced to her class, "maybe I will become like Benazir Bhutto." She was nine years old.

Usman plans to spend the evening with Sadia. They

have a whole night to talk, to just be, and he finds himself retracing the steps of her journey—from that nine-year-old girl to this quiet woman in her *abaya*.

How do you get from there to here? Two years Usman's senior, Sadia was always so much his role model. She was fierce, even confrontational. When she got into the London School of Economics and their mother opposed her acceptance, saying it was too far, too foreign, Sadia called their father and asked simply, "Would it be different if I were a boy?" Tariq had often preached about the emancipation of women in the West; Sadia forced him to be true to his words. And off she went to London.

Her rebellion, and her journey, guided Usman. As Sadia blazed through her first year of college, studying economics and mathematics, Usman prepared for his launch.

But life is tricky, as are best intentions. Tariq and Ayesha were delighted when they found that their daughter could live in the London home of a family they knew well, the Qureshis. Tariq had gone to school with the Qureshi boys, played cricket with them. Sadia would not seem so very far away, living in the London home of people so well known to the Khosas.

After her first year, Sadia came home wearing a head scarf, and there was panic in the family. Tariq and Ayesha were up all night. She had not been raised this way. She was taught to be free from all of that. On each subsequent visit, she had added something, another layer, another article of clothing hiding their beautiful daughter from the world. Sadia embraced the limits, the proscriptions. It made no sense to Tariq and his wife.

Or to Usman, who returns home just before dinner. The tent from yesterday has been taken down. Chairs and tables, packed in their white slipcovers, are being hauled off by the caterer.

There will be other celebrations of Sana and Zain's wedding at the end of December, large, raucous parties with friends, acquaintances, and business associates— an over-the-top Pakistani tradition with origins in the region's feudal past. But tonight life returns to some version of what it has long been as the Khosa children— Sadia, Usman, and Sana—eat a quiet dinner with their mother and father. Afterward Ayesha and Sana go off to work on planning the parties of the coming weeks.

Usman and Sadia retire to the comfortable living room—the formal main room of the house. Tariq joins them, as does Imran, Sadia's husband, who'd been out running errands.

It's a time when Sadia can talk more freely, in the company of her father and brother—*maharim*, or "unmarriageable relations"—who in Islam have special status.

Sadia sits in a severe, straight-backed chair, Imran in one next to her, while the two Khosa men recline across the room, each on a couch, as everyone makes small talk.

Though they talk on the phone, Usman and Sadia have not sat together for nearly two years. Usman's last visit to London was in the spring of 2006, just before his arrest and interrogation. She seems to have moved, since then, even further into the life of faith and piety.

She says she's happy—working as a teacher in an Islamic school. She feels enormously "safe in the world. That's what my faith gives me."

Usman thinks about how Sadia has often described her internal struggle about wearing the *dupatta*—the head scarf—in her early days at the London School of Economics. She was having deep conversations with the religious Qureshi brothers. She began asking herself a new set of questions that were not about human knowledge and learning—the curriculum in her college courses—but about "perfect knowledge," as she would call it: knowledge from God. When she put the head scarf on, it "was like I was struck by lightning. I knew who I was." A fellow student at college, an Asian girl, told her that day that she didn't even know Sadia was

Muslim. It was striking, Sadia said, that "here was the most important thing about me, and she didn't know it." That was January 13, 2000. Two years later, she was in the *abaya*, walking London's streets—a presence in black that would "make people choose how to respond and react to a person of faith," as she tells it. Sadia has said she considers that the start of her new life.

Usman and Tariq want to engage her. The certainty that Sadia embraces, they say, is causing collisions around the world. "The world of Islam," Tariq says, "is clearly going through a battle for its soul. Between extremists and peace-loving Muslims."

"Peace-loving Muslims, like me, who might dress the same as the terrorists," Sadia acknowledges. "Maybe we need to open up more, to explain ourselves."

Imran jumps in, saying that the media distorts "our positions and purpose," and describes how effectively a small radical minority uses symbols, and the media's fixation on image, to hijack the conversation. "I live this reality every day," he says, the reality of language and labels, and because someone like Tariq is called "a moderate, automatically I'm a fundamentalist, and dangerous."

Tariq shakes his head. "But that doesn't mean give up. There must be a place to start. We are in a bad spot now—all of us." Tariq knows firsthand. He spends his days chasing people dressed like Sadia and Imran in western Pakistan, on the borders with the tribal areas and Afghanistan. They look him in the face and laugh. They are brimming with messianic certainty, with a view of the world as black and white, and their time on earth as a brief way station on the path to eternity.

Tariq's battered from it. He sees Sadia in the uniform of his enemy, and Usman on the couch, quiet and tense. Like any father, he tries to stop the battle before it starts, with an appeal: "Islam and Koran is really a very dynamic concept. It is not frozen in the fourteenth century. Therefore, for the twenty-first century and beyond, you have to progress." He talks about the need for interfaith dialogue about what the faiths share. He thinks—he hopes—that "we are moving toward a religion of universality. A religion of humanity. Christianity, Judaism, Islam—call it whatever you wish . . . it is humanity."

Usman listens, appraisingly: admirable, but the "what we share" conversation isn't enough. His sister's in a burka, for God's sake. "Sadia doesn't agree," Usman says tersely. "She thinks Islam is the best and the only way. Right, Sadia?"

She nibbles on the bait. "All religions are from the same sources," she says, "from Allah." Before she can go down the familiar path, saying that Islam, while it affirms the foundation of other religions, is their successor—God's last word—Usman cuts her off.

"If it's really about humanity, an atheist can feel the same way as you," he says, referring back to Tariq's point. "Without any religion you can still be part of the religion of humanism."

Sadia seems to stiffen. Comparing her to an atheist? Usman is clearly ready for battle. So is she.

"Okay, let's try it that way, Usman. I think we can agree there can only be one truth," she says, coolly. "Not two truths. That goes for anything. Right?"

One truth.

"Well, she *is* a mathematician," Imran says, now in Tariq's conciliator role, seeing in advance where this rhetorical line is heading.

Usman sees it, too. "I don't get that," he says, wanting her to go deeper.

"It's only logical," she says. "If something is true, then everything else is not that true thing, thereby not true."

"The premise, Sadia, of having an absolute truth is in itself a very big premise, that some people would challenge," Usman says. "But, fine. So, what if I believe in reincarnation?"

"Okay," Sadia replies. "Take the issue of the afterlife. There can only be one truth about it. I say I have the right answer."

Usman pounces. "No, no, the truth only becomes the truth if one of us finds it out and *proves it*. And there are things that cannot be known by man!"

The war inside Islam's family now rages between the straight-backed chair and the couch, just as it did between al-Bukhari and Avicenna.

Usman and Saida race across the battlefield—charging, retreating, regrouping. It's the real debate, and it's rarely heard. Without propriety or impatience to drive them apart, they can't stop themselves. They are brother and sister, after all.

"So what evidence would be acceptable to you?" Sadia asks, her voice rising. "What would you want to base absolute certainty on?"

Usman pauses, searching. "Evolution, or the evolving of ideas, and experience," and after a minute or two, he settles on "scientific-learning, um, things that can be tested." It sounds small in the wide universe. Usman is referencing the realm—empirical knowledge, the Western canon, experimentation—that Sadia excelled in and then largely abandoned.

"I have evidence," Sadia replies. "The word of God as revealed through the Koran. That's evidence." Then she moves briskly through the structural elegance that differentiates Islam from other religions, how the Koran envelopes the texts of its predecessors, incorporating the story of the Jews and that of the Christians and seal-ing ownership with a divine claim unmatched in the Hebrew Bible or the New Testament: that every word in the Koran comes directly from God. There's a softness now to her voice. "Usman, why would God send anything but the truth, anytime he sends truth to the world? Through Abraham and Moses. Though Jesus. With each prophet he sends the same basic truth, which he then provides, in his own voice—God's words—in Koran. We have to believe in these books or we are not Muslims."

"At some point you start to believe that you're right because you've figured it out. That you have the answer," Usman says, quietly. "And when you do it, you think you can enforce it."

She takes a deep breath, measures her words. "I believe with absolute conviction that anybody who sincerely seeks the truth will find it," she says, in a stab at inclusiveness.

Usman will have none of it. "What about Branch Davidians . . . or some bomber . . . or my Muslim friend who converted to Christianity? It's not you being special. That's the problem. Everyone is starting with your same exact premise."

Then they're all shouting. Sadia, and now Imran, about having the "correct answer," while Usman shouts them down with: "None of us knows!"

Imran, suddenly protective of Sadia, glares at Usman: "How is it possible that I have no doubt in my mind that what I've come upon is correct? No uncertainty. There's not a single belief I have that will make me uncertain."

It's a circular defense. "Fine for you," Usman says dismissively, which seems to enrage Imran.

"You're answer is wrong already, that there's uncertainty," Imran explodes. "I'm here and I'm certain!"

Tariq, watching the fireworks, can barely contain himself. They've been going at it for two hours, his children in pitched combat. "It's the exploitation of religion that is creating destruction," he interjects. "So, therefore, let us start addressing this. Let us converge, let us build bridges."

Imran, who actually works the "interfaith outreach" terrain in London, shrugs. "Okay, here's the problem. We are doing this in the West all the time. We are trying to build the bridges. The problem is, whose standard are we going to adopt to build the bridge on? Whose foundation are we going to use?"

"Don't blame others," Tariq fires back. "They've got to understand, to realize your convictions, your truth, your understanding, your tolerance. You have to make them understand! There's no other way."

Usman looks hard at Imran, who, along with his

brothers, met Sadia not long after she arrived in London and helped guide her transformation. Now Usman teams up with Tariq and starts pressing Imran on the issue of exploitation—of how, in every era, religious fervor has been shaped into a weapon. Imran's response about the "misuse of the text" doesn't impress Usman.

"It's your premise that 'I' know God's will and that absolute knowledge comes from 'my' interpretation—and it *is* an interpretation—of the text," Usman says. "The difference with Islam, how the text is actually God's very words, makes it all that much worse." It's what hands a weapon, Usman says, to bin Laden and to all the bombers.

Imran, double-teamed by Khosas, tries to push back, but after hours of debate, he's boxed in, the remaining uncharted areas are few.

"There's a difference between what Osama bin Laden does and what I do," Imran says, grim and firm, glaring at Usman, and summoning a statement of outrage felt by millions of deeply religious Muslims.

The room is quiet. The night could end in a breach—the kind that might result in a daughter, a sister, going off with her husband and not being seen for a very long time.

"Really, I wasn't suggesting . . ." Usman says, his voice trailing off.

Sadia, sitting in her chair, has been watching the back-and-forth, her large dark eyes, visible through a slit in her veil, darting between father, brother, and husband. They're fighting over the crisis in Islam but also, in a way, over her and the choices she's made.

It is nearly 11:00 p.m. The four have been in this room for nearly four hours.

Sadia looks at Usman, slumped on the couch, emotionally spent.

"You once believed, four or five years ago, that the Koran was absolutely the word of God," she says. "I remember. What happened?"

Too much to explain, he thinks. Where would he start, and where would the story end? He lived. And so did she. But in diverging worlds.

"I guess I should admit that my belief sometimes comes with doubt," Usman says. "That doesn't mean I don't believe. I do. My belief is complemented by doubt. It means I would be more likely to be hesitant, to be shielded from going to the next level. I'd be afraid I might be wrong."

"Well, if someone has come to me and suggested something like that to me, which they have, why haven't I signed on?" Imran says. "Why, then, aren't we terrorists?"

Usman pulls back from the abyss, talking like an

economist. "Each individual choice is meaningful, but even the small percentage that choose violence create a clear trend. Why, for instance, have the last fifty terrorist attacks been with Muslims involved?"

Well, if he can do that, Imran can certainly talk like a doctor. "You seem to see using this premise of doubt, Usman, as being something that is healthy, almost like a preventative measure from you possibly being recruited to that ultimate evil mode."

Usman pauses for a long moment. "I'm saying that doubt enhances human progress."

At this moment, it is fair to say, Usman Khosa becomes an American. He has, after all, carried with him to Lahore the same question that Mary Lisa asked of Ibrahim—"But what do you think?"—and turned it on himself. The war inside of Islam cannot be ignored by the world's greatest power, which happens to have grown up and grown strong on Christianity, Islam's historic—and sometimes violent—competitor. But what can be done? What's clear listening to Sadia and Usman across hours is that many of Islam's dilemmas can be worked out only inside of that ancient religion, by its adherents and combatants. American involvement often tips the scales against best intentions.

The question—"But what do you think?"—carries a golden filament within. It's not revealed in the asking.

It's in the caring. Caring what Usman thinks, and what he says about doubt and progress, reconciling, in his own way, the collisions of piety and ambition, faith and reason, that rage with him.

What happens next is proof of why this American value—a real one, having to do with caring what others really think, rather than imperial concerns of what they can be made to think—will triumph if given the chance. It's infectious.

Sadia finally speaks. Her tone is quizzical, analytical—a voice Usman has not heard in a great while. "What exactly do you mean by 'human progress'?"

Their eyes meet. He—the new American—profoundly cares about what she thinks, and she knows it.

"Shall we figure that out together?" Usman Khosa says, lightly.

From behind the veil, Sadia laughs.

Yes, there are painful struggles boiling inside the family of Islam. And yes, in an era of profound destructive capabilities, they are setting the world on edge.

But what happened in Lahore on this hot night should be duly noted. A brother and sister—one in slacks from Bloomingdale's, the other in a severe *abaya*—managed to search and sweat until they told each they're still part of one body, one shared soul.

———

The airport lounge in Quetta, Pakistan, begins to fill with local politicians, the Bhutto entourage, and a crowd of reporters—from Pakistan's Associated Press to the *Balochistan Times*—with questions about Musharraf's decision to lift the state of emergency. It happened this morning, December 15, in the thick of the election season.

Bhutto sits in front of a tabletop of microphones and fields the kind of questions she can hardly resist—those about Musharraf's declining fortunes. Yes, the lifting of the state of emergency is a positive step, she says, but it's too little too late. "I made a compromise for democracy and it was for the free and fair elections in the country."

Other questions come about the emergency order, giving her the chance to say what lifting it should mean—the release of all the judges and justices, "because their struggle is ours," and all political opponents who were swept up by Musharraf's people. While he's at it, Bhutto adds, he should release the popular head of the region's largest political party—Balochistan National Party chief Sardar Akhtar Jan Mengal—who was jailed by Musharraf in 2006 for allegedly kidnapping two government intelligence operatives. The *Balochistan Times* reporter energetically jots down a few notes.

"Any other questions?" Bhutto asks cheerily.

After so much backroom effort by the United States to engineer an arrangement in Pakistan—to shore up a military strongman whom the White House calls America's most important partner in fighting terror—a strange thing happened in the past month.

Democracy. Not a U.S. version—far from it—but something quite different from what has occurred in recent memory in this part of the world.

In the middle of this drama is, in a way, an improbable character: a well-worn, twice-elected, twice-deposed fixture in the country's long-running soap opera of power and perfidy. She can't quite believe her good fortune.

Bhutto—like other political operators from Islamabad to Washington—is busy hustling to keep up with fast-unfolding events.

When she returned from Dubai on November 3, her status and poll numbers were just a bit above Musharraf's. The confusion following the Karachi bombing, the inability to locate a culprit, and her volley of scattershot accusations kicked up clouds of dust. She was, day by day, getting harder to locate as a martyr.

Musharraf's emergency order prompted a series of rapid moves and countermoves in Pakistan and abroad. On November 4, Rice said that the administration would review its $150 million monthly aid payments to Pakistan, even though she hedged that much of the grant went to counterterrorism efforts that would be unwise to put in jeopardy.

With thousands of political activists detained—and troops securing the streets of every major city—it took until November 5 for the first serious clash to occur. It was two thousand lawyers, confronting police in Lahore, backed up by students and journalists. While the lawyers, again, tried to spur antigovernment activism, Bhutto was trying to consider a next artful move. She announced she'd fly from Karachi to Islamabad for demonstrations, but she and her advisers meanwhile were still working the back channel between the United States and Musharraf, trying to keep her power-sharing deal intact.

Again, it was a play of sunlight and shadow, public events and hushed phone calls, deals made above the table and deals made beneath it. While protests were being crushed and demonstrators swept away by security forces, U.S. diplomats quietly told Bhutto the same thing U.S. ambassador Anne Patterson had said in their first meeting after the Karachi blast: "tone down" any criticisms of Musharraf.

Meanwhile, Musharraf began talking about dates

for taking off his uniform, resigning as army chief, and scheduling elections. Some of his aides suggested that the state of emergency would be short, maybe lasting no more than a few days. U.S. officials started to embrace this line, saying publicly that they thought that would be fine idea.

On November 7, Bhutto did something out of character. She announced that there would be a "long march," from Lahore to Islamabad, the following week.

Discussions about the march actually started two nights before, as she sat up late, exhausted from another day of endless backroom poker with Musharraf and the United States. The lawyers, and especially Chaudhry, were a problem, she said. They owned the "high ground" of principle, which had only been augmented by the previous week's showdown and arrests. While she was spouting democratic rhetoric, she was caught in the deal room—a position in which she came close to mirroring the "say one thing but do another" behavior of the United States, the kind "they're criticized for." Before she slipped off to sleep, she opined with a friend about American civil rights leaders—such as Martin Luther King, Jr.—and what they might have done. The conversation was half in jest—"That's America, this is Pakistan," she said—but talk of the fabled freedom marches in Mississippi spurred the idea of a long march. Just "to see what happens," she decided to announce it.

"If the constitution is not restored immediately, Musharraf doesn't quit his army post and it is not announced that elections are going to be held on time, I ask all political parties to join me in the long march which will end in a sit-in in Islamabad," Bhutto told reporters in Islamabad. "I appeal to the people of Pakistan to come out in groups of three and four with Pakistani flags and if you get arrested, get arrested." She added that "Musharraf can open the doors for negotiation after meeting our demands."

The response by Musharraf's government was befuddled. Information minister Mohammed Ali Durrani said, "Her comment about deadlock seems to be an emotional statement. Every politician knows there is no dead end in political discussions. The long march plan is just a statement and there are many ifs and buts in her statements."

Bhutto saw, to her surprise, that the government was scrambling. It only got worse for them, as the long march drew publicity, especially from the American press and international media. All she had to do, then, was up it one notch, saying she was going to join demonstrators at a forbidden rally in Rawalpindi.

Suddenly, police surrounded her house in Lahore.

The next press briefing she gave was in front of cement blockades and barbed wire. She said any deal with Musharraf was off. She smiled and chatted with the star-struck police officers, told them she knew this wasn't their idea. The photos and video clips streamed across the globe. "Bhutto Under House Arrest." The show of force by Musharraf now seemed to provide context, and a suggestion of official culpability, to the tragic blast in Karachi. A former prime minister imprisoned to stop a march for freedom? Maybe this was about democracy, after all.

Just as easily as Usman and his girlfriend, Sadia, logged on from Dupont Circle and picked up the phone to Lahore, Bhutto's house-bound team tapped into the global news cycles, saw what was unfolding, and started working the phones. Bhutto did round-the-clock interviews from inside her prison. In two days, a South Asian political boss became a global hero. Mayor Richard Daley, under house arrest, reading and rereading "Letter from Birmingham Jail."

Bhutto's poll numbers started to rise, and Musharraf's to drop. This trend continued for the past month. Even after she was released in mid-November, the aftershocks of her house arrest gave momentum to her freedom. But for how long?

Surveys from December on showed Musharraf's approval rating slipping into the twenties, down from 30 percent the previous month.

Bhutto, meanwhile, was swiftly reshaping the political terrain. Her PPP, already the country's largest party, was energized, picking up strength and new members. Even former prime minister Nawaz Sharif—permitted to return in late November by Musharraf, because he thought Sharif would politically dilute Bhutto—let Bhutto lead the charge.

Astonishingly, the United States didn't seem to notice. Bhutto's chief American adviser, Mark Siegel, said later that State Department officials still, by mid-December, felt that "Musharraf's support remained strong. Or maybe they thought he'd just be successful in fixing the elections" scheduled for January 8. "Either way, they didn't seem to recognize what was happening."

What happened, in fact, was that the United States had fallen so far from its basic principles, and was so focused on its best-laid plans and power plays and do-it-but-don't-get-caught cynicism, that it missed an actual contest between democracy's ideals and tyranny's prerogatives.

This contest—this collision—couldn't be much clearer than it is today in Quetta. Bhutto finishes her press conference, saying, "Democracy is the only answer" and the "people are the ones who, finally, need the real power"—things she is, almost against her will, actually starting to believe, as are the growing crowds at her rallies.

She then steps outside into a crossfire of ill will. The most interesting question, at this point—the real question—is who is most threatened by her sudden rise and who, thereby, profits most from her death. The Islamic radicals are now seeing that Bhutto's story, her death-defying democratic rebirth, could energize the moderates in Pakistan, a group that outnumbers the religious radicals by five or six to one and, properly mobilized, could well crush them. Then, there's the entire dictatorial power structure of the country, managed and expanded by Musharraf—and the $11 billion the United States has sent Pakistan since 9/11—which could be swept away by a Bhutto landslide in January.

What's most troubling is this may not be an either/or choice. Official power and religious authority have been cutting deals to run this country since General Zia executed Bhutto's father in 1979 and turned fundamentalist Muslims into colonels and intelligence officers. Thirty years after, the military and intelligence services are rife with radical Islamists, and the Taliban openly run their Afghan war operations from, well, Quetta.

That's the town Bhutto's black Toyota Land Cruiser

is now driving through. It's a crowded, impoverished place, having grown from six hundred thousand to over a million in the past six years, from a steady stream of Afghan refugees.

Ever since a *New York Times* story in January 2007 showed how soldiers fighting against NATO troops in Afghanistan were being openly recruited here from local madrassas—a practice tacitly permitted if not encouraged by the Musharraf government—Quetta has been acknowledged as a key radical stronghold in Pakistan. While reporting the story, the *Times'* Carlotta Gall was beaten up in her hotel room in Quetta by men who identified themselves as members of a special branch of the Pakistani police. Undeterred, other reporters soon began following her line of coverage. Bhutto, who used to visit the city as child with her father, has not been to Quetta in years. A surprising number of people line the streets, some holding signs.

Bhutto, waving to them, sits in the backseat with Naheed Khan, her longtime aide and shadow. A bit older than Bhutto, and a head shorter, Khan—terse and stylishly imperious in her scarf and ever-present sunglasses—was placed in charge of Bhutto's security after the Karachi blast. This means, in essence, that she's always at Bhutto's side, which is one of the most dangerous locations on the planet.

In the front passenger seat is a colonel, a PPP supporter, with a walkie-talkie, who's keeping in touch with the local police and the four white Pakistani Police Security vehicles escorting Bhutto's Land Cruiser—one on each side, one in front, one in back.

The Land Cruiser belongs to the driver, Lashkari Raesani, the local warlord. How do you attempt to protect Bhutto in a place like this? You tap old power, tribal power, with roots that predate centralized government or organized religion. Luskori is head of the Raesani tribe, which has been in a struggle with the area's other major tribe—the Rinds—for as long as anyone can remember. By picking a side, you have half the terrain covered—about the best you can hope for—and you have some clout. Taped to the inside of the windshield is a circle the size of a small melon. It's a cutout of Raesani's face—dark hair, thinning slightly, close-cropped beard, wraparound shades—looking out. The same face that's behind the wheel. Message to all: mess with this vehicle, and you're messing with Raesanis, which, of course, could get you killed.

So, when Raesani gets nervous, everyone in the car gets nervous.

And he's nervous. There have been reports over the past few days of six suicide bombers on their way to Quetta, and Bhutto's their target. Raesani has his

Motorola walkie-talkie pressed to his ear. He grunts something and puts down the device. He looks in the rearview, addressing Bhutto.

Three of the bombers were caught outside of town at a security checkpoint, he says. "Three got through."

He turns quickly, to look directly at her: "Keep going?"

"I don't know, what can we do?" she asks, and looks out the window, at the faces, many of them bearded men or dirty-faced boys, who silently watch the motorcade pass. The car is quiet. Benazir makes an assessment, talking to herself, but out loud, so everyone can hear. "We've heard these reports about these six coming. They've been wanting us to cancel our day here. So, they are real, the bombers, and three got through. But the jammers are good, right?" Those given her last week by government security were broken, but she knows the chief administrator for Quetta—he used to work in Larkana, her home town—and today's jammers seem to be in good condition. Nods all around about the jammers.

"Everyone, keep your eyes open," she says, finally, her voice strained. Everyone does, as the car pushes forward, fitfully, and then slows—some sort of blockage up ahead—and Raesani starts yelling, "Move it, move it," into the walkie-talkie.

Bhutto rubs her brow, and exhales. "A crazy life," she says to Naheed, who just nods, silent, behind her shades.

That's the way the day proceeds, in a state of low-grade, simmering panic. She visits a few elected officials, and a man—an old friend—whose wife has died, before arriving at the main event, an early afternoon speech at Quetta's soccer stadium. After long introductions, where every PPP official, including Raesani—the local party chairman—rises to speak, Bhutto steps up to appreciative cheers.

Those in the crowd—about 1,500—seem grateful they could come at all. The end of the state of emergency doesn't affect the ban Musharraf has imposed on public demonstrations, or more precisely, what's permitted at venues where entry and exit can be controlled. In this case, all those in attendance had to get tickets in advance, at which point they were screened by security personnel; when they entered today, they were screened again, by guards with wands.

More than one hundred officers in plainclothes mill in and among the crowd itself. Another one hundred rim the stadium.

When Bhutto steps up, the anxiety among the security team is palpable. They are working for the region's chief administrator, the man from Bhutto's hometown who has given his word that he'll protect her. Beneath him, though, is his security chief, who actually is overseeing the event; he says, "We're just hoping to get lucky." Standing inconspicuously among the eighty or so local politicians and dignitaries crowded on the bleachers behind the podium, he elaborates: "She knows she's taking a risk, but she has no choice, really. You can't run for office in this country without big, wide-open rallies. Musharraf knows that the level of fear, and all the precautions, work in his favor. People here see that. It's all about a show of power, and showing that Bhutto still doesn't have it, she doesn't have the power to move freely like politicians always have. Of course, it all could be solved if the government moved forcefully to protect her, and had the people in the intelligence services, who are close to the radicals, say, 'Hey, control your people. We want her to live, and we'll come down hard, really hard, on you if you kill her.' But Musharraf won't do that. What's left is us, doing the impossible."

After five minutes of cheering, Bhutto gives her stump speech, ending with the PPP's slogan, the signature call of her father: *Roti, Kapra aur Makan!* It means "Bread, Clothes and Shelter." The new symbolic parallels between the executed father and his imperiled daughter give the old standard a reviving shock, and it draws the loudest cheers of the day.

After the rally, the Toyota Land Cruiser creeps though a dense area of shops and markets.

"Okay, here is good. Stop the car. I want to get out," Bhutto says.

"No, no," Naheed says. "It's foolish."

"Come on," Bhutto says, smiling. "Live some."

In an instant, Benazir's out the door, as police— twenty of them—leap from SUVs and fan out in every direction. She marches across a small courtyard to a fruit stand—Balochistan is Pakistan's fruit-basket and orchard—and picks through the peaches and pomegranates, cherries and apricots. "So, how are things for you?" she asks the merchant, and they make small talk. The police are in close, but the crowd is gathering around—fifty, maybe seventy-five, and growing.

Then she spots them—stacked crates filled with oranges. Maybe ten dozen. "How much for all of them?" she asks the merchant, shoving cash in his hand and shouting, "There are oranges here, for everyone." As the police, passersby, and gawkers all go for the fruit, Bhutto strolls back to her car. Three turns later, the gates to Raesani's compound open, and she's home free.

Bhutto rests in the late afternoon at Raesani's house, a twenty-five-room sprawl of cement surrounded by a private army, about fifty strong, with Kalashnikovs. Raesani is the PPP's chief for Quetta, and later tonight he's hosting a dinner for six hundred in honor of Bhutto.

After a nap, she, Naheed Khan, and her longtime spokesman, Farhatullah Babar, settle in one of the main room's in Raesani's house—a high-ceilinged trophy room with horns on the walls next to mounted sabers, rifles, and machetes.

Bhutto lays a few pillows on the rug and reclines across them, with nuts and raisins, to review where things stand, ponder next steps, and look into the day's controversies and developments.

Babar has a list for her, and runs through it, starting with a situation in India, where the country's national security adviser has just gone on television saying that there was concern in India about Bhutto's leadership and trustworthiness were she to become prime minister. It's a nasty little dust-up, reviving old India-Pakistan grudges, and, by default, affirming Musharraf, who Bhutto thinks is behind it. Babar's already working a counteroffensive, and Bhutto talks to a few Indian journalists, and then to other journalists—from Pakistan and the UK—about "Musharraf trying to steal this election; he needs to be stopped."

She runs through a brief list for reporters of what

must be done, such as replacing the corrupt election commissioners, temporarily suspending the country's mayors—Musharraf's most loyal agents, who are set up to tamper with ballots—and quickly updating voter registration lists. She ends each call with a statement about the power of democracy.

She hangs up and looks over at Khan and Babar. They all know she's in a fix. No one's certain of a way out. The original plan, brokered by the United States, was for her and Musharraf to coexist, but the events of the past month have made that impossible.

"My success and his failure are now the same thing," she says, grimly. "There's no middle ground."

The group sits on the rug, as an hour passes, and engages in a strange kind of fantasy, of what the United States could, and should, do.

They start in the shadows, talking through the kind of backchannel deals that Bhutto is comfortable with. The only way Musharraf and his team will give any ground, Bhutto says, is if the United States does something like "freeze the accounts of the key people around Musharraf. That's the only path to getting any results.

"You've got to hit people where they live, to understand what drives them, and often it's very simple." She runs through a quick disquisition on the back rooms of South Asia, describing the way public officials grab what they can, when they can, to gain access, for themselves and their families, to the world they see shimmering over the horizon in London and Paris and New York. "In America, elected people walk out the door and make tens of millions with Halliburton or dealing with Saudis for some investment bank," she says, nibbling on a pine nut. "Here, they take a cut of the money on its first pass through official hands. That's the way it is in most of the world."

She pauses. "It's important to be honest about all this, because there's opportunity here, to get results." she says, Freeze their accounts, she says, and the men "can't buy jewelry for their mistresses." Khan and Babar begin to laugh. "The wives will say, 'Are you telling me our son has to withdraw from Georgetown, and I won't have the money for what I need?' This is the way you get results. The signal to them is, fine, you love the life in America, the things you can buy, the colleges, or in England? If you want a taste of that, you have to play by certain rules."

She starts running through leading officials, their names and profiles—from top ministers to intelligence chiefs. "And the key is no warning," she adds, "which is what Americans do too much of—always threatening this and that. Then people will hide their money. No, freeze it and say it will be unfrozen if certain things happen."

This, after all, is not a far cry from what the United States threatened to do to her the previous summer, to get her to accede to its terms, to support Musharraf, to behave, to do as she was told. Bhutto didn't know about the NSA intercepts, but a U.S. official let her understand that the United States could, if need be, "constrain her assets," just as she was now suggesting they do to Musharraf.

The darker question in Bhutto's mind: Why hasn't the United States done the same to Musharraf and his team?

"The vice president needs to make the call—he's the only one that Musharraf will respect—and say, 'Here, boy, is how things will work,'" Bhutto says, her eyes flashing. Then she broadens out strategically, beyond her frozen accounts idea, to the larger issue of the manifold threats to her life.

"I'm in a difficult position, but it's not a complicated one. The vice president needs to make a call that I know, from talking to diplomats, he's not made. He needs to call Musharraf and say, 'You hold the cards. If something happens to her, we will hold you responsible.' Why can't he make that call?"

At 8:00 p.m. the dinner is set to begin. Bhutto slips on her sweater and pulls it tight against the chill of the Quetta night. The tent is large, like a circus tent, but wide and low, with space heaters and a long head table near the entrance. The crowd of party regulars, assorted supporters, and friends of Raesani stand and ask Bhutto questions. This goes on for four hours. Finally, slabs of lamb, slow-cooked all day on spits in a wide fire pit, are dropped onto the tables with plate-sized disks of fresh bread.

Bhutto, at the head table, sits among a makeshift delegation from a world in collision—an ancient, toothless Afghan tribesman with seven wives; the head of a beleaguered human rights organization from Lahore; a local politician whose brother is a professor at the University of Virginia; a farmer anxious to sell pistachios in Europe; a mullah who calls himself a moderate but has secretly signed up to assist the Taliban; and Raesani, tonight's host, who is currently on the payroll of at least three intelligence agencies—the American, the British, and the French.

There are no utensils. As great, dripping sides of lamb land, at midnight, on the head table, this cross-cut slice of hungry humanity grabs at the charred flesh, tearing at it with their bare hands. It is delicious.

Bhutto makes it out of the tent in the early morning hours and walks slowly back toward the adobe brick mansion, still thinking of the American vice president. There is clarity to her exhaustion, the end of a long day in which she's spoken nonstop about democracy's promise in an extremist stronghold, all the while being driven through the city with suicide bombers in her midst.

"I took a gamble on democracy, and all I asked of the United States is that they cover my bet," she says, suddenly hard-eyed. "Is that too much to ask? But what I'm beginning to think is that they've made their choice. That's why Cheney's not calling. They decided to go with their favorite dictator rather than this most cherished American ideal. Why would anyone do something like that?"

She looks around. Naheed and Babar are saying goodnight to Raesani. Armed guards are escorting the last stragglers through the gates. The sun will be up in five hours. She'll fly at noon to Karachi and relax for a few days, before resuming her schedule, with stops in Lahore, Islamabad, and, eleven days from now—on December 27—a rally in Rawalpindi.

Now, standing in the gravel driveway of the warlord's house, Bhutto ends up in a place that surprises even her. She's become a believer. "I've talked about these big ideas like democracy and people having real power my whole life. But I don't think I ever really believed in them—not really—until now. If you trust them, they really do work.

"It's sort of amazing."

Ibrahim Frotan sees the lights of Kabul come into view.

He caught a ride this morning in Bamiyan and traveled east twelve hours over Afghanistan's tall central mountains. It's almost midnight on December 19 as the car bumps across potholed highways and into the city's wide dell.

Kabul rests at the bottom of a huge basin, a bowl that held its fill of rubble after American and NATO forces bombed and then stormed the city in late 2001.

When Ibrahim was here in 2003—after his family left Bamiyan and was looking for a place, anyplace, to settle—Kabul was being rebuilt and was just starting to fill up with refugees from across the ravaged country.

Since then, they have come in great waves, nearly doubling the town's population to 3.5 million, as war-tossed Afghans have tried to find a foothold in their relatively safe capital.

Some of those footholds are being carved into the slopes that surround and shelter the city, where people have been busily building makeshift houses. There's no running water on the mountainsides. But there is electricity, wires strung mostly by U.S. troops, which is why the hills are filled with lights; as many lights, it seems to Ibrahim, as in the night sky over Bamiyan.

The car weaves across the quiet city and to the locked gates of a modest compound. Ibrahim gets out and rings the bell, and a guard unlocks the gates.

It's the American Councils' office in Kabul. Inside, Naeem Muhsiny, is waiting.

"About time you made it," he says in Dari. "I'm glad you're here. We have work for you."

Ibrahim laughs and hugs him. It's been a year since their showdown on the phone that night in Denver, when Ibrahim made his gambit to stay in America and Naeem shut him down. Ibrahim is older now and, he likes to think, wiser.

Of course, Naeem will demand proof of such a change. Ibrahim will now have a chance to provide it. The American Councils is launching a new program in language immersion. The idea is to expand the pool of kids the student exchange program draws from by having a few hundred kids a year, ages twelve to fifteen, come to Kabul for two months of intensive courses in English. It's a way to dramatically increase the number of students who ultimately go to America—which private benefactors of the program are anxious for—while keeping the quality and language skills of the selectees high.

A few of the program's alumni who've been to America and are now back have been tapped to be tutors to the youngsters. Ibrahim, with his strong skills in grammar, is one of them. A few weeks from now, he will settle in an empty school that the organization rented for the 220 kids in the language training program, teach classes, and receive a modest wage.

Until then, he'll stay at the American Councils' offices, which have dorm rooms on their second floor. Naeem shows him inside and rouses the head of the Kabul office, Ted Achilles. "Welcome back to the family," Ted says, and throws his arms around Ibrahim. "It's like you never left."

Ted is a character—a divorced, sixty-five-year-old former venture capitalist with a delicate stomach, the refined manners of a museum guide, and a reckless streak to boot. His father was a noted U.S. diplomat who helped create NATO. Ted, aggrieved about America's declining status in the world, came to Afghanistan in 2003, opened the American Councils' office, got his hands on a battered Russian jeep, and began driving the Afghan countryside, interviewing kids. On a trip through the country's remote central mountains in 2005, he met Ibrahim. He liked the kid from the start. Later, he made sure the Washington office didn't ship Ibrahim home from Denver and ruin his grand plans for recruiting more kids like him from Bamiyan.

"So, how are things going up there?" Ted asks.

"Fine, very good," Ibrahim says.

Ted nods, unconvinced. "Well, you're here now, so they're about to get better."

Ibrahim smiles. "I'm grateful you called me," he says, simply, then retires to a small bedroom upstairs and lies in bed, looking up at the ceiling.

Things haven't been going all that well, but for now Ibrahim wants to handle them in his own way. He went to the Provisional Reconstruction Team building in October and waited outside for the PRT civil affairs chief to come out. When he did, they spoke briefly. Ibrahim told him about his trip to America, and the man he met in Niagara Falls. The PRT man told Ibrahim to contact him immediately if he received more threatening letters. That was about a month ago. Nothing else has happened; no more letters.

All of it made Ibrahim jumpy, feeling like he needed to be on the move, starting some new chapter in his life. When he got the call from Naeem about the Kabul job, his thoughts turned to America, and he asked himself, if he were threatened, "What would an American do?" How would an American handle this kind of problem? Ibrahim decided, he says, "that an American would not make a fuss. He'd just keep doing what he was meant to do, what he wanted to do, and not be afraid."

In fact, the phrase "What would an American do?"

is one he often repeats to himself. He wants to return to America, someday, and in the meantime he wants to learn "how to think like an American." Then, he says, "I'll be ready if I get the chance."

Just after sunrise the next morning, mosques across Kabul open their doors to receive finely dressed worshippers. Today, December 20, is the first day of Eid ul-Adha, "The Feast of Sacrifice," one of the major holy days in Islam. It comes just after the Hajj, the annual pilgrimage to Mecca that is every Muslim's once-in-a-lifetime duty. Eid, which lasts four days, commemorates the prophet Ibrahim (Abraham in the Judeo-Christian tradition) and his sacrifice to God. Unlike Jews and Christians—who believe that it was Isaac placed on the sacrificial altar—Muslims believe that it was Ibrahim's first son and the forefather of Muhammad, Isma'il (Ishmael), whom Ibrahim was ordered to sacrifice. During Eid, Muslims sacrifice an animal— usually a sheep, cow, or goat—in honor of Ibrahim's sacrifice, his obedience and devotion to Allah, and the sheep God allowed him to sacrifice in Isma'il's place. The meat from the slaughtered animal is apportioned in thirds, one of which goes, as ritual charity, to the poor and needy.

Those who can afford to will take a day off of work

today. The streets are quieter than usual, the city's residents at mosques and in their houses.

Kabul is a city of locked gates. Every house of any size stands behind a wall protected by armed men. Security is by far the city's predominate business, employing tens of thousands of guards, followed by a huge companion industry of weapon sales, mostly Kalashnikovs. Every man in Kabul who can afford one has one.

The size of this outlay, in a town of limited means, shows where the needle sits on the conventional scale between hope and fear—especially when it comes to breadwinners and their loved ones in those guarded houses. No expense is too great.

A central mission of Hamid Karzai's government—to push that needle into some middle range of guarded optimism—took shape with the building of a five-star hotel, the Serena, which opened to great fanfare in 2005. The city's long-established international hotel, the Ariana, had been taken over by CIA in 2002 for its large Kabul station. Karzai, who strong-armed Arab investors to have this hotel built, put the Serena right across from the seat of Parliament. Both buildings are symbols, but the Serena is the more secure of the two. It has the look of a small walled city, with one high security entrance for vehicles and one for foot traffic. Inside, its marble foyer, two luxury restaurants, and tasteful rooms at three hundred dollars a night—not far off the country's annual family income—are filled with diplomats, war profiteers, and journalists who can demand that news organizations foot the bill as a version of battle pay. To the passing Afghan—who can barely see the hotel's top floors behind twenty-foot cement walls—the Serena is an emblem of Western luxury and the second most impressive fortress in town.

The first is the headquarters of the International Security Assistance Force, or ISAF, the coalition of twenty-six thousand American troops and twenty-eight thousand from forty other countries. ISAF's primary purpose is to project force, and its headquarters needs to be vast. The thousand officers stationed here rarely leave this compound, and almost never without an official reason. Then they move swiftly and purposefully, in military vehicles carrying large visible warning signs—designed to deter suicide bombers—explaining that anyone following within one hundred yards will be fired upon.

What's provocative, though, about the man who runs ISAF—four-star general Dan McNeil, now on his second tour in the country—is that he's thinking less these days about force and its uses and more about the etiology of a commonly used phrase.

"So whose hearts and minds are we talking about?"

he says, sitting in his office in a large colonial-style house at the center of the compound, his boots resting on the coffee table. "I sit here knowing the hearts and minds that I have to accommodate. First, it's the Afghan people. Then it's a group of Afghan—and I won't use the term *elite*—but they are set apart. . . . Then there are some hearts and minds that are just across the border, in *both* directions. Then there is North American hearts and minds . . . and European hearts and minds. They're all different. Their views are all different. Okay, so I often ask myself that rhetorical question: Whose hearts and minds?"

He knows that clarifying this struggle and engaging in it effectively should define America's actions in Afghanistan, Pakistan, and Iraq, and more broadly, its role in the world—though he's not exactly sure of a wise first step.

What is clearer—what he has learned six years into this mission—are the limits of what force can achieve. In an hour-and-a-half-long conversation, he runs through the array of lessons learned: the way NATO forces can defeat the Taliban in any head-to-head battle only to find the enemy popping up elsewhere in whack-a-mole fashion; that new roads and schools are used by insurgents for transportation and shelter; and the fact that Afghanistan's growth into an opium capital—providing

92 percent of the world's supply of all opiates—is fueling the insurgency and undermining U.S. efforts at every turn. That last dilemma doesn't necessarily respond to U.S. aid as you'd expect. Helmand Province—the site of both heavy fighting and heavy opium production—has received more U.S. aid in the past several years than all but four countries in the world. Yet, McNeil says, battles rage in Helmand, and the poppy still grows down there "like Kansas wheat."

The administration announced last week that it would again assess America's role in Afghanistan, which means McNeil, as America's top general in the conflict, has been counseling top U.S. policymakers nonstop—offering advice that will likely go unheeded. The problem is the way American public officials respond to McNeil's "whose hearts and minds" question. They tend to answer it narrowly, with an eye toward a domestic U.S. constituency they know is hungry for good news—the simpler the better.

McNeil, having to hold together a U.S.-led coalition force that includes significant commitments from the British, Germans, French, Italians, and Dutch, is forced to blend competing positions into some sort of consensus, the way the United States has in the past, and may again in the future. That's what makes Afghanistan—one of the world's poorest countries—so different from

Iraq. This conflict predates its counterpart in Iraq and will almost certainly outlive it. Here, a global coalition is involved in a struggle that's both regional, involving Iran and Pakistan, and ideological, deeply entwined in the rise of violent religious extremism. Success will depend on whether the rest of the world, and especially the United States, can muster a meaningful response.

Getting into McNeil's boots has its charms. Not because of the small victories he's presided over—there are very many of those—but rather because of the hard-earned insights he's gained, the best of which come from partnerships that would be untenable in partisan America or within the United States' rather rigid, stick-to-message posture abroad. The best of these unions, he says, is with the United Nations' chief in Afghanistan, a former leftist radical from Germany named Tom Koenigs, whom McNeil—a farm-reared conservative from North Carolina—sits with every two weeks.

McNeil's sixty and will soon enter his fortieth year in a career that's landed him in every major U.S. conflict from Vietnam on. Koenigs, a German intellectual and member of his country's Green Party, has done the same with the UN and assorted human rights groups. The two men's starting points, in fact, couldn't have been much farther apart. Forty years ago, McNeil, then a young soldier, was informed that his brother had been killed when his plane was shot down over Laos. Koenigs in those days was involved in the Vietnam conflict, too. The scion of a Cologne banking family, he gave his inheritance away to what he then considered freedom fighters, including Chilean insurgents and the Vietcong.

In their biweekly encounters—there have been twenty of them—Koenigs and he agree on many things, McNeil says, including how "you have to have security operations in order to do the other things that have to get done." Koenigs, who directs about 1,500 UN employees working on those "other things," such as humanitarian assistance and outreach, has, meanwhile, helped provide perspective on what McNeil considers the essential question—one "that nobody asks" but that he raises with everyone he meets: "What's achievable in Afghanistan?"

McNeil says Koenigs told him, "If we could get them from being the fifth poorest country in the world to the fifteenth, and a functioning government, albeit with corruption—manageable corruption, much like Turkmenistan—it's a success."

That, McNeil says, with undisguised gratitude, was "the most truthful response."

Late that afternoon, Koenigs is puttering around the UN residence, thinking of loose ends. He's retiring in a few days and returning to his wife in Frankfurt. "It's my second wife," he says wryly, "and I've been divorced once. I don't want it to be twice."

Koenigs walks to the kitchen to get a bottle of sweet wine from the industrial refrigerator. The Afghan servants are gone, and all that remains is the security detail. The house is cold and filled with echoes, like an empty hotel. Koenigs says he's "given to reflection" in these few days before his departure. He thinks often "of what we bring to this place, what we'll leave behind," because, he says, "it helps me consider what we're really here for."

The long table where he sits and pours the wine is the closest thing anyone will find to common ground these days in the country. More than anything else, more than the various services and peacekeeping efforts of the mission, "this may be the most important thing" the UN offers, he says, "this table, where people from all the different factions can meet to dine or talk once in a while." That includes him and McNeil—"though we are quite different in every way"—government officials, foreign visitors, and even representatives from the Taliban, whom Koenigs has long advocated engaging in a dialogue. "You lose absolutely nothing sitting down with them, if they are willing to come," he says. "Talking is always better than killing, always, even if you profoundly disagree. What do you lose, exactly?"

One strange twist to those gatherings, Koenigs notes, involves the longtime chief steward of the UN residence, a man named Ahmed. He's been in the job for decades, serving visitors and guests in this room. "But he also is a rather important tribal leader," Koenigs says. "People who know who he is look up startled and say, 'But I should be serving you, sir.'

"Ahmed just smiles. He serves us, but who is serving whom? We come, we think we are in charge. But there are those who have the patience to wait us out. They know that soon enough, in the long history of this country, we will all be gone—the Americans, the British, the Germans."

As the hours pass, and Koenigs talks through his time in Afghanistan and in previous postings, including Kosovo, it becomes clear that his most valued moments have not been in his official capacity, meeting with dignitaries, diplomats, or government ministers. No, they've been simple cross-cultural encounters, the sort that are open to almost anyone with sufficient desire. He talks about one of his staff here—a relative of the tribal leader—who got an education through the UN, and "how she came back, renegotiated her contract and, well, she killed me" in the negotiation. There's "a development question tucked in there," he laughs, "about whether we really want them to become like us. Because when they do, when they pick up abilities, they will challenge you—and it'll cost you. Of course, it's worth the price."

Another memory strikes him, a story that changed his thinking about Sharia law, something he'd long reviled, despite knowing little about it beyond a few punitive practices. A driver who worked for him, a young Muslim man from Bangladesh, fell asleep at the wheel one night and killed another young man in a car accident. The driver would have faced a manslaughter charge, but instead, he went to the father of the dead man and said, "We are all Muslims. This is my fault. I fell asleep. I am profoundly sorry. Is there a way we can resolve this?"

"The father was an old man," Koenigs explains. "He said, 'My son, who you killed, supported our family. Now you will be my son and you will support our family.'"

Koenigs decided not to alert the authorities about the manslaughter, and instead called upon Sharia judges, who codified the arrangement. "Sharia law, I've learned, is much more victim-oriented, rather than the way it is in modern legal systems, where it's the state versus the perpetrator in a courtroom, and the victim is nowhere to be seen," Koenigs says. "My driver has a United Nations salary. Rather than jail and ruin, he now supports this family, and they consider him one of their own.

"I still consider the Western legal system superior, but look at how I was forced to learn from this improbable experience, how much you can learn, even as an old man."

He pauses, pours another glass of wine. "We've become too sure of ourselves, we in the successful West," he says. "We stopped listening. We lost our capacity to be surprised, to be taught.

"The solutions really lie within us, we with the strength, with capacities. The key is for us to show the Afghans—who have much greater knowledge of their situation and much more at stake—that they can teach us and help us understand what we can do for them that will be helpful. The country is filled with men, many of them young men, of great promise. When we listen to them, they're strengthened, and then they learn to guide us.

"But," he says with a sigh, "to do this, you must be humble. And we are anything but humble."

Koenigs gets up and walks around the dark mansion, making sure its rooms are locked and its lights off before he retires to his quarters. He stops at another table—a more formal one, in a darkened dining room— and something strikes him, something that happened a few months before.

A group of tribal elders from just across the border in Pakistan came to see him. They sat here, a dozen men in bright robes, religious men. "They wanted to know if I could redraw the border so that they would be on the Afghan side. I told them I certainly didn't have power to do that. But I asked them why would they want such a thing. The oldest of them, an uneducated man, but very bright, said, 'We cannot read, and our grandchildren cannot read. We think it is time they were able to read.' I replied, 'So, they can receive religious education?' 'No, no,' he said. 'They already know all the religion they need. We want to them to learn all the other things so they can be part of the world.' "

"That's why they came all that way to see me," Koenigs recounts. "They live in Pakistan's tribal regions, where there is no law, no government, no schools They thought if they were in Afghanistan, at least the kids would have a school. They were just old men, like me, worried about their grandkids."

As he rolls the story over in his head, Koenigs—the UN's special envoy, racing back to Frankfurt to save his marriage—smiles, an idea taking shape inside him.

"If you look hard enough, you know, you can see the face of all of us."

Ibrahim Frotan is doing clerical work that Naeem assigned him, to prepare for the language program, but his mind is on a desktop computer—a nice one, a Dell—that he saw in the lounge upstairs at the American Councils' office.

It has a good dial-up connection, but people are usually on it. This much is clear: if he doesn't get some quality time on the Internet soon, he'll lose his mind.

He feels like he's been waiting for months. There's a small Internet café in Bamiyan, with an old IBM desktop. But it's useless. You have to wait hours to get to the computer, and when you do, the slow dial-up connection often cuts out. People download programs, and then share them around the village. But IM, or e-mail, or trolling websites? No way.

That's the thing he misses most about America, he's decided. The glorious Internet. He'd always loved sitting in front of computers. He could spend his life at the keyboard; it was a point of pride. He won a citation for being "best with computers" at his school when he was fourteen, and a teacher had him apply to a program bringing surplus computers to Afghanistan for students, to help them go online. The problem was with the application. Ibrahim's family had not used a last name for several generations, and the application seemed to require one. Ibrahim considered a few, trying them out on his friends, before he settled on Frotan. In Dari, it means "humble." Ibrahim liked the word. The clerics at the local mosque said often that being humble was a virtue "that opened the heart and let the soul breathe." Ibrahim remembered that. The first time he wrote his new surname was on that application. The computers never came. But his desires—to get a computer, to go online—are what gave him his name.

He thinks about all this as he files applications from the kids selected for the language program. First he went to the United States, and now he'll guide others. He feels a sense of renewal, and possibility, a little like the way he felt on his last journey, his bag packed with the pin-striped suit from his uncle and gifts for Ann Petrila and Michael and Ben. That's why he is wearing the gray University of Denver sweatshirt Ann bought him, with the red and white letters. The hard lessons he learned in Denver helped him succeed in Pennsylvania, and all those experiences, he's certain, will help him succeed now.

Naeem, who's been busy running errands, stops by Ibrahim's desk in the midafternoon, and they chat for a little while. "So, tell me more about how things went for you in Kane," he says, pulling up a chair. "From what I understand, things worked out pretty well."

Happy to get relief from the filing, Ibrahim proceeds to tell Naeem some of the stories from Kane. It's thoroughly liberating. He's told his mother and father all about it, and showed them pictures, but it was difficult to describe so much of it. There were so few shared reference points.

Naeem's eager to hear every detail, including, and it seems especially, about Ibrahim's friend Jillian. Ibrahim had told his parents nothing of her, afraid they would misconstrue it. Now he can be completely honest talking to Naeem, an older guy with experience in both worlds.

He tells Naeem the whole story, and gives special attention to the months after Jillian told him about her daughter, which were the best times for Ibrahim. Everything was easy and natural, at school and at home. He engaged his teachers. He worked to express himself in new ways through art projects. He and Jillian would sit every day in the cafeteria, waiting for the buses, and Ibrahim came to understand everything, including the reason Jillian looked askance at a boy who sat across the cafeteria with his friends. It was the baby's father. He looked like a tough guy—big and loud, wearing a dark jacket—and he was always eyeing Ibrahim. The boy left school in late April. People said he dropped out.

Around that time, Ibrahim and Mary Lisa talked in the den one night after dinner. There were posters and fliers all over the school about something called "prom." Jillian said she was on the committee working on the decorations. Ibrahim asked Mr. Johnson, the gym teacher, about it, and then trolled the Internet to learn more. It was a very big celebration apparently, to which the boys asked the girls and where everyone dressed up. There was so much online about the ceremony that it was hard to know where to start.

He told Mary Lisa he was going to ask Jillian to the prom. Mary Lisa said he shouldn't wait, because someone else might ask her. "Don't you break that girl's heart, Ibrahim," she said. Ibrahim laughed. She'd said that a few times before. "No, no, Mom," he said. "I would never do that to Jillian."

Now that they were on the subject, though, Ibrahim could get all his questions answered. Tom was working that night, the dishes were done, and Mary Lisa seemed to know everything there was to know about proms.

"You know, Ibrahim, you're going to have to dance at the prom," Mary Lisa said, suddenly very intent.

"I know how to dance," Ibrahim said, at which point he jumped up from the couch and began to dance, his hands high and hips out, to show her.

"No, no," she said, shaking her head. "You have to dance *with* the girl, touching her."

"Okay, okay," Ibrahim said, wanting Mary Lisa to know he'd set aside certain religious prohibitions. "Jillian and I have touched each other. It's not a problem."

Mary Lisa just sat there for a minute, not saying anything.

"What music do you listen to, that's slow?" she asked. Ibrahim found something, and Mary Lisa then showed him how to hold her hand up high and put his other hand on her hip.

"This is the way I did it at my prom," she said, sort of quietly. "These things don't ever change much." Then the music came on, and they started to waltz.

From that night on, Ibrahim called Mary Lisa his "prom coach." She said he was a quick learner.

Two months later, he was in the pin-striped suit. With his monthly stipend from American Councils, Ibrahim bought Jillian a necklace. He gave it to her that afternoon at school, with a corsage he'd bought her. He was sure he saw her eyes well up.

That night she was wearing everything he bought her, looking beautiful, in a delicate light blue dress.

It was a prom, like every prom.

The air was full of anticipation, as everyone arrived at the gymnasium, decorated to the fullest, with a great colored cardboard arch through which the couples passed, and a mirror ball—the girls all looking at one another's dresses, the boys watching the girls watch one another.

Ibrahim and Jillian danced for hours, laughing and sweating and dancing more. They were picked by the teachers as the winning couple in a dance competition—an ecstatic moment.

Jillian then spotted a boy—a friend, who was distraught because his date had left him—and decided to dance one dance with him. Ibrahim said, fine, he'd dance with a girl, a tall dark-haired girl from his gym class.

They were both looking at each other as they danced with other people and, for an instant, it was like they were strangers.

An hour later, when they reconnected in the foyer, Ibrahim said something about buying a soda for Jillian—that she must be thirsty—and she said that he didn't need to buy her things. But he bought them anyway, two Pepsis, which they carried around, undrunk, like lead weights, until Tom came by to drive them to an after-party and they deposited the sodas in the cup holders of his car. At the after-party they watched movies and didn't say much to each other.

It was just before dawn when Tom picked them up

and drove them toward Jillian's house. They sat silently, in the backseat, neither one sure what to do. There had been so much buildup to the big night, and things had been going so well. Ibrahim just sat there, feeling that he wanted to say something—to apologize for whatever he'd done to upset her—but he couldn't summon any words that might work. Jillian thought of all the times they'd laughed together, and how gentle he was, and the story about the boy on the island, and the girl who finds him. That was her. She was the one who rescued him, or she wanted to be.

The car pulled up to her house.

She reached over and put her hand on his. "Why don't you come in? We can crash here together."

He was surprised she'd asked, delighted, and he started to move. Then it hit him.

Dogs. Jillian's mom was a dog lover. They had five dogs in the house, big ones. Sleeping with dogs? It was like a hand had reached up from some dark past and clutched his throat.

"No, I don't think so," he said, softly. "But thank you for asking."

Jillian just shook her head and got out of the car.

"Don't forget your Pepsi," she said, and slammed the door.

Things were never the same after that. They chatted. It was cordial. But a spell had been broken. Ibrahim was soon to leave. Mary Lisa and Tom took him on a few trips, including the one to Niagara Falls, before graduation. He stood with the class, in his cap and gown, and was soon back in Washington—for a week of "reentry" counseling from American Councils—and then on a plane to Kabul.

Of course, Naeem was present for that last part. Still, he listens to it all, rapt.

"You certainly did learn a lot in America," he says. "I'm glad we didn't send you home."

"Me, too," says Ibrahim. "Very glad."

"Now, get back to work," Naeem says, smiling.

After dinner, Ibrahim finally gets his chance. There are a few other people staying in the bedrooms upstairs, but none of them is around. The lounge is free.

The computer awaits.

He signs on to his Yahoo account, and sees that his e-mail in-box is overflowing. There are six months of messages, gigabytes of spam.

He scrolls through page after page, and then back to the top, to today.

It is Friday, December 21, the second day of Eid, and he has notes wishing him a happy feast and peace from other kids in the YES program, friends Ibrahim can't wait to respond to.

He reads some e-mails and sends Eid notes back. It's like he's alive again. He could do this for three days straight. But there's something he's been wanting do. That first. He opens a screen, and begins to type:

My Dear families:

Hope you are all fine and doing well. I am very sorry that I could not send you email since a long time, I am very sorry about that. Now I am sending this message from the American Councils office. Now I am helping and teaching English in the English Imersian Program in Kabul. I miss you so much my dear families. I would like to see and talk to u face to face like before.

Please keep in touch and take care

Have fun and have a good one.
Sincerely:
Mohammad Ibrahim Frotan
(Bamyani)

I wish the best time for you my dear I hope hear you soon. :)

He taps out the e-mail addresses of Ann Petrila and Mary Lisa Gustafson, types SALLAM AND EID MUBARAK TO ALL OF U!!! on the subject line, and hits Send.

Far away, a girl is sitting at her computer.

A contact she hasn't seen in nine months pops up on the "friends" list of her instant messenger. She catches her breath.

There's no time to think.

IBRAHIM?

There's a pause. She watches the space beneath her message in the IM dialogue box.

JILLIAN, IT'S ME. I'M ALIVE!

I'VE MISSED U SO MUCH IBRAHIM, she types.

Then the white space fills, again.

I MISS U 2 JILLIAN. EVERY MINUTE OF EVERY DAY

This story, at its finish, now issues its challenge, subtle and smiling, to each observer. How, with a world in need, will you fill the white space in your life?

One thing is now clear: the boy is no longer on an island, fearing monsters, and the girl is not so certain who saved whom.

Chapter 4
Truth and Reconciliation

Benazir Bhutto always disliked Rawalpindi, the garrison city.

The military—with its history of conducting affairs with blithe autonomy and seizing power from time to time in authoritarian coups—is stationed here.

It's also a place with a history of violence. Liaquat Ali Khan, Pakistan's first prime minister, was assassinated here, in 1951. And this was the place, in 1979, where Bhutto's father, Zulfikar Ali Bhutto, was hanged. Bhutto's already outlived her father by three years; two more and she'll eclipse Khan.

All that is long past, and there's a campaign to win. A rally has been planned. It's been announced that she'll speak, and the rest of the day's schedule is tight: an early meeting with Hamid Karzai, the Afghan president, another with European election monitors, and then, after the rally, dinner with two members of the U.S. Congress.

This last meeting is particularly important. Bhutto's close colleague Latif Khosa, the Pakistani senator and first cousin of Tariq Khosa, has just finished writing up a 160-page dossier with hard proof that Musharraf's government agencies are planning to rig the upcoming parliamentary election. Bhutto intends to give copies of the dossier to U.S. senator Arlen Specter and Representative Patrick J. Kennedy when she dines with them tonight.

First there is a speech to attend to, and over lunch in her Islamabad home, Bhutto reviews her remarks. Her nerves have settled since last night, when her husband, Asif, called from Dubai to tell her he was nervous. Now, looking up from her notes to glance at the rugged mountains off in the distance, she feels calm.

Then it's time, at a quarter to four, to drive the eight miles to Rawalpindi. Bhutto and a handful of PPP officials convoy down in two armored SUVs. A few hours before, four supporters of another Musharraf rival, Nawaz Sharif, were gunned down on the Rawalpindi streets. Everyone's concerned. But the group's there before they know it, pulling into Liaquat Bagh (Liaquat Park) the very place where Prime Minister Khan was assassinated more than half a century before, at the beginning of this country's history.

When it comes time to speak, Bhutto lays aside the speech she was tinkering with over lunch. She looks out from beneath her white headscarf, over Liaquat Bagh and the thousands of supporters filling its grounds.

"Wake up, my brothers!" she shouts, hands clamped to the lectern. "This country faces great dangers. This is your country! My country! We have to save it."

A cheer rises from the crowd, and it's thunderous, like a river rushing through a valley. She gives one of her best speeches. When she finishes, she is buoyant as she slips down the stadium's back stairs to the waiting Land Cruiser. She wonders where her security detail is—they're nowhere to be found—but she's quickly in the Cruiser, which begins to push through the teeming supporters. They're chanting her name, screaming, ebullient. She asks that the sunroof be opened. She wants to acknowledge them, and rises into the evening air to look out over the crowd. She lifts her hand, smiling brightly, and begins to wave.

An hour later, crowds are laying siege to Rawalpindi General Hospital, smashing its windows and breaking its doors. Inside, doctors are trying to keep her alive.

The nation is stunned, disbelieving. Cries of "Musharraf is a murderer" echo from Rawalpindi to Karachi, where guns are discharged angrily into the air. The country is soon in flames. Most Pakistanis just cry.

At 6:16 p.m., Benazir Bhutto is declared dead.

In the coming days, there will be arguments about whether Bhutto was shot or simply hit her head as a result of the forceful shockwaves from the suicide bomb. She was buried quickly; there was no autopsy.

Eventually, CIA will affirm the opinion of Pakistan's police that al Qaeda and its affiliate, the Pakistani militant Baitullah Mehsud, were behind the killing. Oddly, neither of them take credit for it, though that is their common practice. Normally, Al Qaeda and its Taliban partners kill as a function of policy and rhetorical strategy.

It is difficult to know how Bhutto died, or who was behind her killing, and that may forever be the case. There are so many ways to die in Pakistan.

In the last months of her life, though, Bhutto was doing something that is quite rare for widely known global actors: she was evolving, in public. The near-death experience in Karachi, the house arrests, the surprising way Musharraf's actions seemed to spur a latent self-recognition in Bhutto that she might have a wider historic role than simply that of a politician returning to power, all caused the blood to quicken among Pakistanis and, beyond them, the global community. Whether al Qaeda was behind her death, there is no doubt that bin Laden and Zawahiri—men who acutely understand the power of story—recognized this, and understood that Bhutto was creating a powerful counterpoint to bin Laden's saga of violence and salvation.

In the house of Raesani, the warlord, eleven days before her death, Bhutto looked up at a picture on the wall of her crafty and brilliant father, who was outmaneuvered in the end by an Islamist general.

"My father would say, keep going, keep fighting, but it's more complicated than just that, than one battle after another. You have to find a way to stop the fighting."

Sitting on pillows, letting her mind wander, she talked then about what she'd learned, as a Muslim woman who'd led men "by making them feel they are in charge, and then guiding them to do what you want."

She laughs at this. "You see, they think they're saving you, and you think you're saving them. That's where the trouble starts. Someone says, 'I saved you, now here's what I want.' And it's the same with big countries and little ones, religious leaders and their followers, even husbands and wives. When things really work, though, it's because people realize that this is a lie, that, really, we all save one another. It's the way of the world.

Things work out for the best when everyone makes it, together, when we manage to save each other."

It is a tragedy that American leaders, so convinced of their tactical skills and the shrewdness of their power plays, would ignore the unexpected—Bhutto's evolution—and not see how she might turn into the next great narrative of this period, capturing the imagination far and wide and turning it away from destructive certainty. The United States should have done whatever was necessary—including sending over a few hundred Secret Service agents or pulling together a small international security team—to make sure Bhutto lived to see Election Day. It was a matter of will. Cheney never made the call Bhutto was hoping for. He and the president, once again, trusted illegitimate power over stated principles. They went with Musharraf.

Yet it is nonetheless hopeful that Bhutto—who'd known compromise and corruption and lust for power—would end up where she did.

What was happening, as she came down from on high and pushed, finally, into the crowds, buying them oranges and holding babies—letting herself bleed into them, and them into her—was that she was becoming a representative character, giving herself over to the sensations and struggles and insights that are common to all.

It's something Gandhi and King felt, and maybe Robert Kennedy in his last days: that the giver receives, the receiver gives, and then they become indistinguishable. It's what transformed those men. Bhutto wasn't their peer, but she saw it.

Everyone saves everyone. No one is owed.

When people in the United States and abroad talk about the need for America—alone, now, as the world's great power—to rediscover its capacity for moral leadership, they're talking about actions rooted in the kind of transforming insights Bhutto stumbles across in the last days of her life.

If there is a sense of urgency about America managing this, it is because the past few years are offering a glimpse of what a leaderless, and increasingly savage, future may look like. That's why you find so many of the characters in this book searching for moral energy, both in their own lives and in the life of their country. There are natural links—morality, after all, deals with matters of right and wrong, codes of conduct for individuals, which are created by and define a society. The effort, in a democracy, certainly, is to apply these basic human codes—honesty, compassion, respect for others—to the activities of a nation. When the gap between the individual and the collective becomes too great, it creates a sense of dislocation and prompts someone like Candace Gorman to dissolve into "oneness" with Ghizzawi and the lost inmate in the next cell, or Rolf Mowatt-Larssen to spend late nights thinking about the conflict between secrecy and personal authenticity, or Wendy Chamberlin to pass hours considering how "doing the right thing means you ask nothing in return." Chamberlin, for one, arrives at conclusions similar to Bhutto's, where a personal truth, about the power of selfless giving, carrying no debts, helps her understand the difference between the Marshall Plan—a pinnacle of moral leadership— and United States' recent history of transactional foreign engagements, where actions are all but designed to create debt and obligations that serve various stated, or concealed, American interests.

The great teacher of this in recent times is, of course, Gandhi, who continued mightily down the path of merging his personal morality—and behavior—with a universal morality, asserting that if it is right for you, it is right for me, and that I will not ask of anyone, or any country, something I will not do, even more stridently, myself. At the end of his life, Gandhi—maybe the century's greatest moral leader—all but expanded into a sense of oneness with the wide world, walking barefoot until his feet blistered, living among the poor, starving himself to stay connected to humanity's suffering.

In April 1947, having just set in motion a chain of events that would lead to the creation of Pakistan and his own assassination, Gandhi, in a recorded address, recounts how "a friend asked yesterday, did I believe in one world? Of course, I believe in one world. And how can I possibly do otherwise." But then Gandhi offered a kernel of strategic advice, of how power—moral power—is housed in the valley of despair and human need where he was spending so many of his last days. "You can redeliver that message now in this age of democracy, in the age of awakening of the poorest of the poor."

Of course, the awakening of the poorest of the poor, and their heirs, has continued apace in the ensuing years—they are increasingly connected to one another, can see the world as it is, and are rapidly arming themselves. And they are sentient, and often quite clear-eyed in their struggles, as Gandhi knew.

Which raises the issue of America's most effective way to now engage with that world. There are more and more voices by early 2008 who seem to be allying with the characters, and their journeys, and calling for initiatives—much like the ones Chamberlin mentions—in which the United States will commit to simply making certain every twenty-first-century citizen of the world has fresh water, electricity, basic medical care, and a school. A cross section of Americans—young people, engineers, construction workers—will build these things, and the United States will ask nothing in return. That last part, of course, is where a complex personal truth about selfless giving creates the moral standard that guides a nation.

Attempting such feats requires not only an ability to control power's self-protective, tactical impulses but also, as the shrewd British terror expert David Omand says, to "embrace ever increasing standards of ethics and accountability." What Omand—who has been in pitched combat with jihadists who take vows of poverty and chastity before they attack symbols of secular power—is saying is that shoring up a nation's "ethics and accountability" is crucial to success in this era's struggle.

That's why, before America's most notable actors, the president and his team, walk off the world stage, there is a quiet, steady effort under way to face, finally, basic issues of ethics and accountability.

The full story of Tahir Jahil Habbush is part of that effort—an effort to embrace difficult, often humbling truths on the path to reconciliation and the reclaiming of moral energy, the source—as Gandhi would advise—of the transforming power.

After all, as people know from their own lives, truth matters.

The story of the United States and the Iraqi intelligence chief ends with hubris and deception.

But it starts, though, with a modest, rather humbling, search for truth. The key, in the fall of 2002, was that liberating admission "I don't know."

"We didn't really know whether Saddam had weapons or not. There was a lot of stuff assembled; everyone knew that. Everyone was sure he had them. But it was more assumption than hard facts. This was about a last attempt to find out what we could."

That's Rob Richer. He was at the center of this mess— a final effort by the clandestine intelligence community to unearth something verifiable on the issue of Iraq and its weapons of mass destruction.

In early December 2002, Richer held a meeting at the CIA office in London. He had just been appointed head of the agency's Near East Division. He called all twenty-two of CIA's station chiefs throughout the troubled Middle East to London, and let loose on them.

"He just said, flat out, that all the station chiefs had failed—that we were a disgrace," one of them recalled. "The basic idea was that we had no sources, no chits to call in, no reasonable leverage to get any of the needed information, and that everyone has to work from this moment forward like they could be fired tomorrow."

Abuse wasn't Richer's goal; it was preparation. He

then brought in the man he considered just about the best intelligence officer covering the region: Michael Shipster, the Mideast intelligence chief for SIS, British intelligence.

Shipster ran through a daylong intelligence primer on more than a dozen countries, with special emphasis on the links between Iraq and its neighbors. It was a working session, a sharing of sources and insights that went late into the night.

The next day, Richer and Shipster huddled. They'd known each other for years, having walked similar paths: both had been stationed in the Middle East and Persian Gulf. In the 1990s, both had overseen operations in Russia for their respective services.

Now Richer leveled with Shipster. The United States had plenty of assumptions about Iraq and its weapons, but still, somehow, lacked the hard, irrefutable evidence it needed to make a case for war. Richer had unloaded on his chiefs because he viewed this as a failure of intelligence gathering, of clandestine operations.

Shipster agreed. More important, he had a plan.

He had a source inside Iraq whom the British had worked with before over the years: Habbush.

The Iraqi first emerged as a public figure in the early '90s, when he was the governor of Dhi Qar, a province in southern Iraq that historically had given

Saddam's regime trouble. By the mid-'90s, Habbush had moved into the Ministry of the Interior, where he was undersecretary for security affairs, and worked as the country's police chief prior to taking over as head of Iraqi intelligence in 1999. His predecessor, Rafi Dahham al-Tikriti, died under mysterious circumstances, most likely killed on orders from Saddam. Habbush took over as head of the Iraqi intelligence service, or the Mukhabarat, as it's commonly known. Like many of Saddam's senior officials, he had blood on his hands. Saddam respected those who would kill on command. Habbush was such a man.

But he could be reasoned with, Shipster said, and he knew how to get to him. The heads of both the UK and U.S. intelligence services—Dearlove and Tenet—were apprised, and the two Mideast chiefs went to work.

Habbush would need to get out of Baghdad, meet Shipster in some hidden locale, talk through all of the allegations about Saddam's weapons, and then return, undetected, to Baghdad.

Richer knew of one country bordering Iraq that would be ideal for such a high-risk meeting: Jordan.

He placed a call to Saad Khayr, the head of Jordan's General Intelligence Department, or GID. It is another one of Richer's cozy relationships. Richer'd helped Khayr get his job in 1999, and the Jordanian chief is godfather to one of Richer's sons.

Khayr cleared the way, and a secret meeting was set for early January 2003.

Amman was the perfect place for the two men to meet. The U.S. relationship with the Jordanians is close, and no one understands the Iraqis better than the Jordanians, who share a border with Iraq and have many citizens of Iraqi descent. In Jordan, every aspect of the high-level secret meetings could be monitored by the United States, as could Habbush's passage back and forth to Baghdad, five hundred miles to the east.

While the United States was preparing its case for war—with Bush's 2003 State of the Union still weeks away and Colin Powell's presentation before the UN not until February—the secret back channel to Saddam's regime was officially opened.

Both the White House and Downing Street watched intently.

The first meeting in Amman was mostly to set ground rules. This was the start of a dialogue. Habbush suggested that Saddam knew he was coming, but of course, Saddam had no way of knowing all that Habbush and Shipster might discuss. Also, Habbush said that if the United States invaded, he wanted to be taken care of with passage to safety.

Shipster told Habbush it was not a matter of *if*. The Americans were serious, ready to invade.

Habbush then told Shipster that if they did, they'd find no WMD. Iraq had no such weapons.

The first preliminary report was passed up the ladders, on both sides of the Atlantic.

The reaction inside the White House was one of surprise, then skepticism, from the president on down. After being told that Habbush had said there were no WMD, Bush was frustrated. "Why don't they ask him to give us something we can use to help us make our case?" he told an aide.

Bush understood from the start—as did CIA—that responsibility for Habbush would eventually be handed over to the United States. Terms of that deal had not been worked out. That would be for Richer and CIA to handle later on.

Bush, Cheney, and top aides to the vice president wanted Habbush, in essence, to earn his passage. The United States was working furiously on "the case." It needed damning disclosures, not the Iraqi intelligence chief—who was given the code name "George"—saying there were no WMD.

Meanwhile, Richer was bringing the heads of the Iraq Operations Group—a diverse team of nearly two hundred CIA operatives whose sole focus was Iraq—into the fold. The team, which fell under Richer's Near East Division, included operatives who'd been working on Saddam and the Iraqi terrain since the mid-1980s.

"This was a huge opportunity," recalls John Maguire, one of two men who oversaw the IOG and who'd been working clandestine operations on the country for thirteen years. "It was a window into Saddam that doesn't have an external filter on it. You have an ability to directly reach the leadership."

Maguire and the other man atop the IOG, who's still undercover, started meeting almost daily with Ian McCredie, the head of British intelligence in Washington. From the first meeting forward, they began to plan ways to use "George" and the secret back channel to its fullest. Beyond the issue of what Habbush was saying about the absence of WMD, there were discussions about plans ranging from using Habbush to draw Saddam into a negotiation over exile to having Habbush or others assassinate the Iraqi leader.

Questions about the value and use of Habbush raged through the upper reaches of the government. Everyone had an agenda. Often those agendas collided.

The dominating issue remained how Habbush was undercutting America's stated justification for the war. On that score, "everybody was terrified that we are being set up and will be made to look like patsies," said A. B. "Buzzy" Krongard, CIA's number three official during this period. "Twenty percent are saying he's the 'real McCoy' and he's being truthful. Twenty percent are saying it's D and D: denial and deception. The rest are saying, Jeez, I don't know. Nobody wants to put their name on it and say I'm taking responsibility for it, that this is the real thing and there are no WMD."

Under pressure from the White House to vet Habbush, some CIA managers were saying that Habbush should be viewed as credible only if he handed over the identities of some of his key foreign agents.

The Iraq Operations Group bosses were outraged. "Great. So we're saying to Habbush if you want us to believe what you say about WMD, you have to sacrifice some humans for us," says one CIA boss involved in the debates. "They had no understanding about how this works. It's a window. It goes both ways. You work it forcefully and creatively. You can put anything you want through that window, and Saddam will have to react."

Shipster, who had been the British intelligence chief for Washington in the early '90s, "knew our system very well, and he wasn't going to jerk Habbush out of his socks with these stupid loyalty tests or any of this buffoonery coming from Washington," says Maguire. "He kept the channel open and running."

The weekly meetings in Amman continued through late January and into early February. There were also phone calls between the two men.

Shipster repeatedly pressed Habbush on the issue of WMD.

"The problem Habbush faced," Richer says, "was having to prove a negative to testify to his credibility. That's a trap that we pushed him into, and then fell into ourselves. But he couldn't come up with the proof that the weapons weren't there. And we all figured they were there. We didn't listen to him or act particularly creatively in trying to figure out a way he could provide hard evidence for what he was saying."

Ultimately, Habbush could not offer proof that weapons that didn't exist, didn't exist.

By early February, the British were ready to deliver a report to the Americans.

Richard Dearlove flew to Washington to present the report to Tenet. The British had worked hard on this. The meetings between Habbush and Shipster were one of the world's best kept secrets. Dearlove wanted to deliver their conclusions in person.

The report stated that according to Habbush, Saddam had ended his nuclear program in 1991, the same year he destroyed his chemical weapons stockpile. Iraq had no intention, Habbush said, of restarting either program.

As for biological weapons, Habbush had significant credibility—that program was run by the Iraqi intelligence service. He said that since the destruction of the Al Hakam biological weapons facility in 1996, there was no biological weapons program.

All of this turned out to be true.

As soon as Tenet digested the report, he knew it was trouble. He called in Richer.

"They're not going to like this downtown," Tenet said. "Downtown," of course, meant the White House.

Richer tried to reassure him, saying it was a complex picture about WMD and "there's a lot we know, though there are still questions."

Tenet briefed the president and Condi Rice. Rice sat through the briefing and read the report. She looked at Tenet, for whom she had little respect. "What the hell are we supposed to do with this?" she asked. Tenet said "what to do" was a question for the White House. He tried to place the report in context, talked about how much of the already assembled "case" for WMD contradicted Habbush, what Habbush's motives might be, whether Saddam knew of the extent of what Habbush had reported to Shipster.

One thing Saddam would never have sanctioned was his intelligence chief's description of his boss's state of mind.

Habbush said Hussein was isolated and diminished,

and had little idea what was going on inside his own country. Saddam was worried about the Iranians, and other countries in the region, finding out that he had no WMD. Beyond that, he thought the Americans were bluffing. "According to Habbush, Saddam's view was, 'Why would the Americans want to take over this country? It would be a nightmare. It made no sense,'" Richer recalled. "What Habbush told us about the mind of Hussein—that was some of the most valuable reporting."

The White House then buried the Habbush report. They instructed the British that they were no longer interested in keeping the channel open.

The British were upset, but felt that with this mission they'd done everything they could do. Richer casts the transatlantic debate incisively: "The Brits wanted to avoid war—which was what was driving them," he said. "Bush wanted to go to war in Iraq from the very first days he was in office. Nothing was going to stop that."

Over at the Iraq Operations Group, there was despair. The consensus was that due to the White House's conflicted feelings over how Habbush undercut the WMD case, they'd shut a window that might have been strategically valuable.

"We were disappointed," says Maguire. "We looked

at it as an enormous opportunity, another one, being missed." Continuing to work the Habbush back channel "was a way of throwing so much confusion into Saddam's operations that his ability to direct forces to counter the invasion would be really limited and we might not lose as many men. We tried to explain this to people downtown," he says. By twisting information or creating new incentives for Habbush, the United States could "get a gun battle going" between factions of Saddam's power structure. Maguire says he and other IOG chiefs explained to policy makers that people in Saddam's command structure "are impulsive and paranoid to a level that is difficult for any American to appreciate," that the "addictive drug of choice is Valium" for the jumpy, sleepless men around Hussein, and that "we should be pouring gasoline on the family friction points and to try to cause a rupture before we show up."

"Wouldn't it be nice if we could drive to Baghdad for a meeting with someone who'd just overthrown Saddam," Maguire says, "rather than fight our way there."

Had Habbush given the president "something we can use" on the WMD case, the back channel might have been kept open. It didn't work out that way.

In Baghdad, Habbush was anxious to know the White

House's response. He received word through Shipster that the United States didn't want to know any of what he had gone to great pains to reveal. A few weeks later, near the beginning of March, he met in Baghdad with a Lebanese American businessman named Imad Hage, who was trying to act as an intermediary between the Iraqis and some Pentagon officials he'd met. Habbush reiterated that there were no WMD.

According to the reporting of *The New York Times*'s James Risen, Hage then said to Habbush, "Why don't you tell this to the Bush administration?" Habbush replied cryptically, "We have talks with people."

By then, such encounters had become a lost cause.

"The bottom line is that the Brits did something we're not very good at, which is to set up this sort of back channel and get a good solid report," Krongard says. "The White House was all but saying, 'This guy Habbush is telling us things we don't want to hear. Make this go away.'"

Now the only question left for Washington was: What to do with Habbush?

On March 19, the United States began its "shock and awe" campaign, as American bombers unloaded thousands of tons of explosives on Baghdad.

Habbush was ready. He slipped out of Baghdad with

the help of U.S. intelligence and into Amman, Jordan, where he'd had his meetings with Shipster.

Battles, meanwhile, raged in Iraq. Saddam's troops fought against the U.S. military, though the resistance was not what the United States had expected. On April 9, Saddam's statue toppled before cheering crowds, as U.S. troops poured through Baghdad.

There was much to do. U.S. soldiers, weapons inspectors, and CIA agents fanned out across the country. They checked site after site that had been listed as locations of WMD.

By early summer, it started to become clear to U.S. intelligence officials that—as some had suspected in January—Habbush was right. There were no WMD in Iraq.

"Everyone was holding their breath," says Maguire, "saying that I hope Habbush doesn't pop up on the screen."

At around this time, CIA worked out specifics of its arrangement "to resettle" the former Iraqi intelligence chief. The agency agreed to pay him $5 million out of CIA's accounts. The amount, Richer says, was based "on an assessment of the information he provided and what he might provide in the future."

By that time, Habbush had found his place on Bush's famous blacklist of Iraqi war criminals, which the president fashioned as a deck of playing cards. Of course, the top card, Saddam Hussein, was the ace of spades. As leading Iraqi officials were killed or captured, Bush would mark an *X* across the face of the corresponding card. The deck became gift shop chum, Internet fodder, and a collectors' item for soldiers. Habbush, who was officially listed as missing by the United States, was the sixteenth card in the deck.

He was the Jack of Diamonds.

Rice and Cheney were "pleased that we had him, that he was now our guy," says Richer. "But they didn't do much with him."

As summer progressed, and the absence of WMD started causing more and more tension inside the White House, Habbush, in fact, became the last person the White House wanted to talk to.

"We should have been spending twenty-four hours a day for weeks talking to this guy," says one intelligence official involved in the Habbush case. "He was their intelligence chief. He once ran the police for the entire country. He knew Iraq inside and out. It was stupid not to talk to him."

The public started to find out something had gone terribly awry on July 6, when former diplomat Joseph C. Wilson published his opinion piece in *The New York Times* entitled "What I Didn't Find in Africa," which undercut the White House's claim that Hussein was attempting to buy yellowcake uranium in Niger. On July 11, Tenet took the blame for letting the president make this claim—the famous sixteen words—in his 2003 State of the Union Address, as though it all had been a problem of communication between CIA and the White House. On July 14, columnist Robert Novak revealed the name of Wilson's wife, Valerie Plame, exposing her identity as a CIA agent and starting a chain of events that eventually resulted in the conviction of the vice president's top aide, Lewis I. "Scooter" Libby.

On October 2, after three months of exhaustive search, the Iraq Survey Group issued its interim report to Congress that weapons inspectors had yet to find any evidence of WMD. "We have not yet found stocks of weapons," said David Kay, the group's chief. "But we are not yet at the point where we can say definitively either that such weapon stocks do not exist, or that they existed before the war."

That was when Habbush finally got his promised cash. "The beginning of the fiscal year is October first," says Krongard—a former CEO of Alex. Brown & Sons, the brokerage house—who ran all the business operations of CIA. "That's when everyone gets paid."

By then, the White House had finally thought of a way to use Habbush.

In late September, Tenet returned from a meeting at the White House with instructions for CIA.

He called Richer into his office. "George said something like, 'Well, Marine, I've got a job for you, though you may not like it,'" Richer recalls.

The White House had concocted a fake letter from Habbush to Saddam, backdated to July 1, 2001. It said that 9/11 ringleader Mohammed Atta had actually trained for his mission in Iraq—thus showing, finally, that there was an operational link between Saddam and al Qaeda, something the Vice President's Office had been pressing CIA to prove since 9/11 as a justification to invade Iraq. There is no link. The letter also mentioned suspicious shipments to Iraq from Niger set up with al Qaeda's assistance. The idea was to take the letter to Habbush and have him transcribe it in his own neat handwriting on a piece of Iraq government stationery, to make it look legitimate. CIA would then take the finished product to Baghdad and have someone release it to the media.

Even five years later, Richer remembers looking down at the creamy White House stationery on which the assignment was written. "The guys from the Vice

President's Office were just barraging us in this period with one thing after another: run down this lead, find out about that. It was nonstop. Of course, this was different. This was creating a deception."

Richer passed the directive down the chain, to the Iraq Operations Group.

Maguire, in late September, was preparing to return to Baghdad to help run CIA's station—the largest foreign station the agency had set up since the Vietnam War. It had been a frustrating summer. The tendency of the White House to ignore advice it didn't want to hear—advice that contradicted its willed certainty, political judgments, or rigid message strategies—was spreading quickly to core operations in Iraq. This became especially clear, Maguire recalls, in August, when he—at that point head of the IOG—presented to policy makers a key intelligence discovery: Saddam's Baghdad Security Plan, the regime's prewar blueprint for how to withdraw key forces as the Americans invaded and then eventually launch an insurgency. "We'd just pushed Saddam's Baathists out of power. We didn't kill them. This showed how they were going to come back with a vengeance." The administration, at that point, didn't want to hear the word *insurgency*. They would not allow it to be used in most meetings or in public statements. The Iraq Operations Group's chiefs, who were regularly advising policy makers in the White House, were lambasted. The report was dismissed. By fall, as the insurgency started to take shape exactly the way it was foretold in the Baghdad Security Plan, Maguire was ready to get away from Washington and some of its practices of denial and deception.

A few days before he left, Richer took him aside and briefed him on the Iraq Operations Group's next assignment: the Habbush letter.

"When it was discussed with me I just thought it was incredible," Maguire recalls. "A *box-checking* of all outstanding issues in one letter, from one guy."

While the assignment was clear enough, Maguire said he told Richer that the logistics would need to be reworked. "Habbush was not going to write, sign, or say anything publicly—even though we resettled him—which was going to sentence his family to death," says Maguire. "Habbush knows there'll be an insurgency and there'll be payback for anyone who's suspected of cooperating with the Americans before the war or just after it started.

"That's the part we fault George on. He should have told them right off the bat that this won't work because Habbush won't sign it," Maguire says. "Tenet obviously can't just say no to a direct order from the White House—he's got to go through the motions—but that would have given him a way out."

But Tenet, who had no background in clandestine operations, didn't engineer a "way out." Instead, he passed an order from the White House down the ranks to Richer, who then passed it to IOG for relatively simple execution. A handwritten letter, with Habbush's name on it, would be fashioned by CIA, Maguire said, "and then hand-carried by a CIA agent to Bagdad" for dissemination.

Shouldn't be a problem. Habbush, in hiding, money in hand, will be inclined to keep quiet.

Maguire was happy to leave the logistics of letter writing and delivery to his successor. He was off to Baghdad—the insurgency was kicking up, and he had a good idea about how it would unfold.

The year 2003 was now coming to a close. Behind the curtain of secrecy, it had already been a uniquely eventful year.

The White House first ignored the Iraqi intelligence chief's accurate disclosure that there were no weapons of mass destruction in Iraq—intelligence they received in plenty of time to stop an invasion. They secretly re-settled him in Jordan, paid him $5 million—which one could argue was hush money—and then used his captive status to help deceive the world about one of the era's most crushing truths: that America had gone to war under false pretenses.

In this case, there is one final complication.

Under a 1991 amendment to the statutes that in 1947 created CIA and that govern its actions, there is a passage that reads, "No covert action may be conducted which is intended to influence United States political processes, public opinion, policies, or media."

The operation created by the White House and passed to the CIA seems inconsistent with those statutory requirements.

It is not the sort of offense, such as assault or burglary, that carries specific penalties, for example, a fine or jail time.

It is much broader than that. It pertains to the White House's knowingly misusing an arm of government, the sort of thing generally taken up in impeachment proceedings.

Con Coughlin, of *The Daily Telegraph* of London, is a journalist whom the Bush administration thinks very highly of. They also like his newspaper. *The Daily Telegraph* is one of the largest British papers, with a daily circulation of nearly nine hundred thousand. Like its owner in 2003, Conrad Black, the paper is decidedly conservative.

Coughlin had reported extensively from Baghdad, and had often written stories about Saddam's vast WMD arsenal. In the run-up to the war, he published *Saddam: King of Terror*, a fall 2002 book released under separate titles in the United States and the UK, illustrating Saddam's atrocities. Coughlin, who continued to file dispatches from Iraq after the invasion of coalition forces, was a favorite of neoconservatives in the U.S. government.

As 2003 came to a close, Con Coughlin received a present.

On a trip to Baghdad in late November, he met with Ayad Allawi, the Iraqi politician and exile who was part of the Interim Governing Council and would soon become Iraq's first head of government. Prior to the war, as the founder of the Iraqi National Accord, Allawi had passed information about Iraq's trove of WMD to British intelligence. Now he had an extraordinary document, the Habbush letter, to pass on to Coughlin.

After one of Allawi's aides slipped the document to Coughlin, the journalist made his way back to England to do follow-up reporting. Allawi told Coughlin he was convinced the letter was authentic. Coughlin said he talked to a former CIA official who confirmed that this was what a memo from Habbush would look like.

By the second week of December, Coughlin was putting finishing touches on a 1,500-word blockbuster for *The Sunday Telegraph*, the paper's biggest story of the year. He wrote that the "the tantalising detail provided in the intelligence document uncovered by Iraq's interim government suggests that Atta's involvement with Iraqi intelligence may well have been far deeper than has hitherto been acknowledged."

Then came the story's key paragraphs:

Written in the neat, precise hand of Tahir Jalil Habbush al-Tikriti, the former head of the Iraqi Intelligence Service (IIS) and one of the few named in the US government's pack of cards of most-wanted Iraqis not to have been apprehended, the personal memo to Saddam is signed by Habbush in distinctive green ink.

Headed simply "Intelligence Items," and dated July 1, 2001, it is addressed: "To the President of the Ba'ath Revolution Party and President of the Republic, may God protect you." The first paragraph states that "Mohammed Atta, an Egyptian national, came with Abu Ammer (an Arabic nom-de-guerre—his real identity is unknown) and we hosted him in Abu Nidal's house at al-Dora under our direct supervision. We arranged a work programme for him for three days with a team dedicated to working with him . . . He displayed extraordinary effort and showed a firm commitment to lead the team which will be responsible for attacking the targets that we have agreed to destroy." There is nothing in the document that provides any clue to the identity of the "targets", although Iraqi officials say it is a coded reference to the September 11 attacks.

The second item contains a report of how Iraqi intelligence, helped by "a small team from the al-Qaeda organisation", arranged for an (unspecified) shipment from Niger to reach Baghdad by way of Libya and Syria. Iraqi officials believe this is a reference to the controversial shipments of uranium ore Iraq acquired from Niger to aid Saddam in his efforts to develop an atom bomb, although there is no explicit reference in the document to this.

Habbush writes that the successful completion of the shipment was "the fruit of your excellent secret meeting with Bashir al-Asad (the Syrian president) on the Iraqi-Syrian border", and concludes: "May God protect you and save you to all Arab nations."

While it is almost impossible to ascertain whether or not the document is legitimate or a clever fake, Iraqi officials working for the interim government are convinced of its authenticity, even though they decline to reveal where and how they obtained it. "It is not important how we found it," said a senior Iraqi security official. "The important thing is that we did find it and the information it contains."

A leading member of Iraq's governing council, who asked not to be named, said he was convinced of the document's authenticity. "There are people who are working with us who used to work with Habbush who are convinced that it is his handwriting and signature. We are uncovering evidence all the time of Saddam's dealings with al-Qaeda, and this document shows the extent of the old regime's involvement with the international terrorist network."

Near the end of the article, this paragraph appeared:

Intelligence experts point out that a memo such as that written by Habbush would of necessity be vague and short. "Trained intelligence officers hate putting anything down in writing," said one former CIA officer. "You never know where it might turn up."

The story—to be published in *The Sunday Telegraph* on the morning of December 14—was leaked a day early and started heating up the global new cycles. It was first picked up by wire services in America on Saturday evening, around dinnertime.

A few hours later, even bigger news exploded. Saddam Hussein had been captured in a spider hole near Tikrit.

The next morning, on *Meet the Press*, Tim Russert began the program by announcing the news about Saddam's capture and then cutting to Tom Brokaw in New York, who had been working on the story through the night.

The news of Saddam's capture was a perfect engine to carry along the companion headline about the Habbush letter. After talking to various NBC correspondents about how U.S. troops had found a beleaguered, bearded Hussein, Brokaw cut live to a special report:

BROKAW: Let's go to London now to Con Coughlin who is a biographer of Saddam Hussein and has an article in today's *Sunday Telegraph* about Saddam Hussein, his regime, and Mohamed Atta, who was the lead terrorist in the attacks of 9/11.

First of all, Mr. Coughlin, did you expect that Saddam Hussein would be taken in such a meek and utterly degraded way, that he would be hiding like a rat in a hole with just a few aides around him?

COUGHLIN: Well, I thought that Saddam would be found in this kind of location. That's for sure. And I was in Baghdad fairly recently and was told that basically Saddam was living in a hole somewhere. And that turned out to be entirely accurate. But what I didn't expect was to see Saddam being cooperative and talkative. I thought Saddam would put up more resistance, more of a fight, and, basically, that he would never allow himself to be taken. Because, basically, by appearing like this with the coalition, the whole myth of Saddam Hussein, that he's invested in so heavily, in the last 35 years, has been destroyed in five minutes.

BROKAW: And tell us about the article that you have today in the *Sunday Telegraph* about Mohamed Atta and any connections that he may have had to the Iraqi regime of Saddam Hussein.

COUGHLIN: Well, this is an intriguing story, Tom. I mean, basically, when I was in Baghdad, I picked up a document that was given to me by a senior member of the Iraqi interim government. It's an intelligence document written by the then-head of Iraqi intelligence, Habbush to Saddam. It's dated the 1st of July, 2001, and it's basically a memo saying that Mohamed Atta has successfully completed a training course at the house of Abu Nidal, the infamous Palestinian terrorist, who, of course, was killed by Saddam a couple of months later. Now this is really concrete proof that al-Qaeda was working with Saddam . . . This is a document, and I've had it authenticated. This is the handwriting of the head of Iraqi intelligence, Habbush, is one of the few people still at large who is in the pack of cards. And it basically says that Atta was in Baghdad being trained under Saddam's guidance prior to the 9/11 attack. It's a very explosive development, Tom.

BROKAW: Thank you very much, Con Coughlin, in London this morning.

Coughlin and *The Telegraph* kept pushing the story forward, adding another disclosure in a companion piece that Allawi had "said the document was genuine.

'We are uncovering evidence all the time of Saddam's involvement with al-Qaeda,' Mr. Allawi said. 'But this is the most compelling piece of evidence that we have found so far. It shows that not only did Saddam have contacts with al-Qaeda, he had contact with those responsible for the September 11 attacks.' "

Over the next few days, the Habbush letter continued to be featured prominently in the United States and across the globe.

Fox's Bill O'Reilly trumpeted the story Sunday night on *The O'Reilly Factor*, talking breathlessly about details of the story and exhorting, "Now, if this is true, that blows the lid off al Qaeda–Saddam."

Nodding along, O'Reilly's terrorism analyst, Evan Kohlmann, added that "we know that the intelligence official who supposedly authored this letter is still on the run, and is now considered one of the top officials still on the run from U.S. authorities."

The story was discussed the next morning on NBC's *Today* show, then on CNN. In his column in *The New York Times* on Monday, William Safire wrote that on Saturday night, "I stuffed myself on lamb chops and potato pancakes at a holiday party at the home of Don and Joyce Rumsfeld. Along with other media bigfeet, I chatted up Rummy and C.I.A. chief George Tenet, both of whom were in on the secret of the capture of

Saddam a few hours before. Neither man even hinted at a thing. So much for being a Washington Insider."

At the end of the column, Safire mentioned the sorts of things Saddam would most likely not admit under interrogation: "Example: Dr. Ayad Allawi, an Iraqi leader long considered reliable by intelligence agencies, told Britain's *Daily Telegraph* last week that a memo has been found from Saddam's secret police chief to the dictator dated July 1, 2001, reporting that the veteran terrorist Abu Nidal had been training one Mohamed Atta in Baghdad."

When asked about the contents of the Habbush letter Tuesday night on CNN's *Lou Dobbs,* Kansas Republican senator Pat Roberts, chairman of the Senate's Select Committee on Intelligence, said, "We have been aware of that story. I'm not going to say it's an old story, but it's a story we've been aware of. That is part of our inquiry. This is a new development. We want to make sure that we have access to that commentary and to the reports. I can't really get into any more specifics, except to say that we are following that up."

It wasn't until Wednesday that *Newsweek*'s prize-winning reporters Michael Isikoff and Mark Hosenball offered the first challenge to the Habbush letter, citing the FBI's evidence of Atta's whereabouts during this period and saying that in Baghdad there was a bustling market for forged documents. Still, news programs, and especially Bill O'Reilly—with nearly 3 million viewers—continued to push the story.

Meanwhile, at CIA headquarters, Buzzy Krongard watched what was unfolding with horror.

"I heard rumors around the building that it was ours, a CIA operation, and I just shuddered," Krongard said. "Basically, my position was 'I didn't want to know.' Because, if I knew, I'd have to resign."

Like Krongard, Rob Richer understands the statutory complications at hand.

"The mission," Richer says, "was intended to affect Iraqi public opinion. To try to affect U.S. public opinion would be illegal."

Maguire, who left CIA in 2005, says it clearly "was more for consumption in the U.S. than Iraq. He adds that "the whole WMD thing was irrelevant to most Iraqis."

But Maguire, who regularly represented CIA in meetings with Bush and Cheney, sees the letter as part of a "foolish arrogance" that dominated the administration's actions from the start. He recalls a particularly revealing moment—a meeting at the Pentagon in spring of 2002—"that says it all about what went wrong." It was a large meeting, about twenty people, in a conference room at the Pentagon, filled with senior officials from the State Department, CIA, NSC, and the Defense Department, including Rumsfeld, to discuss phases three and four of the Iraq campaign—post-invasion and stabilization/rebuilding. While General George Casey, Jr., who was handling plans for those phases for Rumsfeld, was going through his presentation, officials from State and CIA were raising objections that "this would be the toughest part," as Maguire recalls, and "that some of us had been to Iraq and you couldn't pull this off by flashing up a couple of PowerPoints." Rumsfeld, who told CIA and State that their involvement in the planning process was no longer needed, ended the meeting, abruptly, with a signature statement: "We will impose our reality on them."

Looking back, Maguire shakes his head. "Incredible arrogance, in the face of facts and reality, from start to finish, and even making us create fabrications like that Habbush letter."

Richer sees it all more simply, more personally.

"It's called lying."

As the open-source world races forward, the U.S. government's reliance on official lies has left it in a strange stasis over much of the past decade, where the most pressing issues of America's role in the world can't be fairly assessed or debated.

Seminal issues are handled inside of government on two tracks, as they might be by the accountant of a troubled company who is keeping two sets of books. There's a false ledger for the public auditors, and a real one, which only a few top executives are privy to.

This divide has presented untenable challenges for designated public auditors, from congressional oversight committees to the media's most skilled interrogator, the late Tim Russert.

Since taking over NBC's signature show in 1991, Russert had been one of Washington's best corrective measures. He and his staff would read everything in every noteworthy publication and then compare it to previous statements by a guest. With the camera rolling, he'd bring left and right hand together, as the guest squirmed. Then Russert would launch forward with follow-up questions.

An interview in March 2006 with Secretary of State Condoleezza Rice, however, is particularly revealing about how the government's two ledgers really worked in these tumultuous years.

Early in the show, Russert presses Rice about Iran. The Iranians have been increasingly confrontational toward the United States in the past two years. The United States seems to be running out of options. Russert asks if it's "the policy of our government that Iran will not be allowed to develop a nuclear weapon."

It's a setup. Rice says, yes, that it's the position of the United States that "Iran cannot be allowed to develop a nuclear weapon." Russert counters with a quote from a *New York Times* story: an anonymous administration official says the consensus in the White House is actually that there's no way to stop Iran from producing a nuclear arsenal and we "just hope we can delay the day by 10 or 20 years."

Rice chafes, says this is wrong, and stresses that the administration is banking on U.S.-led international pressure on Iran, diplomatic measures, UN Security Council resolutions, and the like. The bottom line is that the United States will act as a firm and honest broker leading the world. Eventually, the Iranians will see the light.

On the secret ledger, however, the bottom line looks very different. The defining event in Iranian-U.S. relations actually occurred in the summer of 2003. It has colored everything since.

That summer, the United States snubbed Iran. The rebuff surprised certain senior officials in the Iranian government; it didn't seem to make any sense. The Iraq invasion, for all its coming complications and costs, had created one hoped-for outcome. It had seemed to change the behavior of the region's most powerful country, Iran.

Having 150,000 U.S. troops next door made a difference. At this time, the Iranians suspended the part of their nuclear weapons program that was most overt and incriminating—their project to weaponize enriched uranium.

They were ready to talk—about the al Qaeda cell within their borders discussing the purchase of Russian suitcase nukes, and about other issues, including assistance with the U.S. mission in Afghanistan.

They sent a message saying as much to the British, who'd kept up relations with Tehran.

The British set up a meeting in Geneva between CIA and an Iranian delegation. It was to be a starting point for talks—a back channel—that might develop in many positive ways. Much like the channel opened by Shipster and Habbush, this one would provide an inroad to another country's leadership.

The administration decided to send Jim Pavitt, CIA's director of operations. Pavitt, who many former CIA officials criticize as ineffectual, all but volunteered, and decided he wanted to take one of the planes George Tenet used—sort of an Air Force One for CIA. Pavitt, who'd spent most of the previous few years fighting battles in Langley, didn't consider how official flights leaving Andrews Air Force Base can get backed up. Tenet's plane sat on the tarmac, behind other government planes, for hours.

Which meant Pavitt arrived late in Geneva.

Then he went to the wrong hotel.

Soon the British were on the line, frantic, wondering, "Where the hell is Pavitt?" according to one U.S. official involved in the matter. The Iranians were livid. They felt disrespected.

A month later, the snubbing would become official policy. Talks were cut off with Tehran. In fact, they'd never really gotten started.

The message from America could not have been clearer, said one senior intelligence official involved in the matter: "Even after the Iranians crawled to us on their bellies, we said, 'Get lost.' They got the message. We didn't want to talk. We wanted them gone. Clear enough?"

Top U.S. officials, especially at the Defense Department, had a slogan in those days: "Real men go to Tehran." A stated goal of the Iraq war—that the U.S. invasion would shape the behavior of other rogue states—seems to have been another in a series of false rationales.

From that point forward, the Iranians acted with a kind of entrepreneurial zeal—harboring al Qaeda, sending troops and weapons across the border to undermine the United States in Iraq, defying U.S. demands that they cease uranium enrichment, even sending arms to support the Taliban. The United States began its efforts secretly to undermine the Iranian regime and discover a rationale for a direct attack.

Much as Rolf Mowatt-Larssen said, thinking about the blank stare from Bush on this very issue: snubbing Iran is "a World War Three scenario." Rolf didn't know about the Pavitt meeting. All he'd heard, like others in the upper reaches of the government, was that Cheney had shut down the talks with Iran.

Without public knowledge of the 2003 snub, a meaningful exchange, on *Meet the Press* or elsewhere, cannot occur. Nothing really makes sense.

The real bottom line: by placing so much on its secret ledger, the administration profoundly altered basic democratic ideals of accountability and informed consent.

But that's not all that happened this Sunday morning in March 2006. Russert saved his biggest surprise for later in the show. He grilled Rice about the first public disclosures—reported by *NBC News* a few days before—about Naji Sabri, the Iraqi foreign minister. After discussing how it was clear that Sabri, through an intermediary, had passed information to the U.S. government that there were no WMD, Russert said, testily, this is "a far cry from what the American people were told" in the run-up to war.

"Of course, Tim," Rice responded, "this was a single source among multiple sources."

Five months later, in August 2006, Michael Shipster made a trip to America. He was planning to leave British intelligence after nearly thirty years and wanted to see some of his old American colleagues.

Shipster is a dashing character, his friends agree, and well liked. "He is sort of the best the British system produces," said an American official who worked with him. "Looks a little like Peter O'Toole, erudite in all sorts of way, rides motorcycles, has a taste for adventure and favors tweed."

His father, Colonel John Shipster, served in the British Army for four decades, in campaigns that stretched from leading Indian troops in World War II to support of the Americans in Vietnam.

The Shipster family subscribes to old-world standards of service and professionalism, some of which were bruised, Shipter felt, by the Habbush matter.

During his visit, he met with friends from CIA at Washington's Palm Restaurant, a pricey downtown steakhouse. They talked late into the night about the world they knew and how it had changed.

There were enough knowledgeable people at the table that they could sort through some of the denials and deceptions that had led up to the Iraq war. They had access to the secret ledger. Sabri's disclosures—out now for several months—troubled the British, Shipster said. The memo they'd received from Washington about Sabri in late 2002 said essentially the opposite of what had recently been revealed.

Sabri's disclosures would have added weight to everything that Habbush had so forcefully asserted, Shipster told a few of those present. What's more, had Sabri's entreaties not been dismissed in the fall of 2002, he might easily have acted as a second channel, not as valuable as the intelligence chief, but there to check or confirm the information Habbush had offered.

"Shipster was upset," said one of the CIA officials at the dinner. "Think about it. Because we conned the British, our closest ally, about Sabri, they couldn't place in proper context the incredibly valuable channel they'd set up" with Habbush.

If the British had known everything that *both* men

were saying, Shipster told his dinner companions, "We never would have gone to war."

In 2003, Tahir Habbush settled in a large section of Amman, Jordan, called "Little Baghdad," and stayed put. That part of Amman is home to as many as seven hundred thousand people of Iraqi descent, many of them recent refugees.

Former U.S. intelligence officials, who've kept track of his status, say Habbush lives quietly, in a modest though tasteful home behind a high gate. He receives a few select visitors and old friends for tea. They sit on a patio in his garden.

In 2005, Iraq issued a $1 million reward for his capture; he's wanted for war crimes. He is still listed as missing by the United States.

John Maguire says Habbush is protected by the Jordanians and "the old Bedouin custom of receiving a visitor as a friend, and making sure they are safe." Maguire says he thinks Habbush would be tough to find in Amman—"there are a lot of people there to protect him"—and that there may even be a role for Habbush in Iraq someday. "If there's a political solution where the Sunnis ever get a real foothold in the government, maybe they'll bring him back. Who can say?"

Rob Richer is back from another adventure.

It's March of 2008, and the Harleys have just roared across South Africa.

This time, he and King Abdullah have brought along a prince from Bahrain and a few other young royals from the Middle East. The Wild Hogs live.

Richer has survived and even prospered. The Blackwater incident in Iraq in September 2007 has largely blown over. The company is currently negotiating a set of new contracts with the U.S. government. Richer's intelligence firm is in great demand. As the United States struggles with its clandestine capabilities, more tasks are being farmed out to private contractors.

In fact, at day's end, Richer's friendship with Abdullah is precisely the kind of relationship the United States could use quite a bit more of. It's a cross-border bond between two men, from starkly different backgrounds, who talk to each other with utmost frankness. Abdullah is arguably the most important honest broker left in the troubled Middle East. Richer supports him with a level of candor and cunning you can't buy at any price. Beyond that, they're best friends. They do things for each other and ask nothing in return.

It is precisely because of that friendship, Harleys and all, that Shipster and Habbush had a safe place to meet in one of the most important encounters of this period.

Reflecting now on the Habbush matter, Richer struggles to put it into context. He was not the source of any of the key information, so he now feels comfortable placing it all in a broader context.

"The legacy of this era is how the U.S. political system manipulated intelligence and didn't play it straight with the American people," he says. "The story of Habbush, from beginning to end, is maybe the most dramatic single emblem of that."

Rolf Mowatt-Larssen's latest disruptive idea is taking shape.

This one, finally, might be his exit strategy.

He's deconstructing secrecy—that's the root of it. Having not been able to tell people what he did for much of his professional life, not having a business card until ten years ago, he feels he knows the concept, both its benefits and perils.

As he's become a bit more public in the months since Thanksgiving, moving from one speech to the next, he's found that the discussions in daylight—at think tanks, institutes, universities—are more productive, more informed in many ways, than the ones he has had in some closed rooms inside the government. It's a sensation shared in the past few years by countless officials with top security clearances. Once they start having wide-open talks on their topics of interest and expertise, they start to realize how little secrets are worth.

Driving home one night in April 2008, he felt the weight of years' worth of "how" questions—how to find out what other nations may be doing with their loose uranium, how to assess the gap between what they are saying publicly and doing privately, how to know when, or if, you've detected a network.

That's when he started asking "why" questions.

Namely, why should this whole area of terrorists and nuclear weapons be handled in secret? What, specifically, is the value of secrecy in such a matter, beyond good old-fashioned CYA—cover your derriere—the government's rationale for much of what it keeps sealed away?

In fact, no nations refuse to acknowledge the peril of a terrorist obtaining the materials to build a nuclear weapon. So you put everything in sunlight—with plenty of wide and growing international cooperation—and force countries to be true to their word.

Over the coming weeks, he stitches together an outline, a framework of offices in countries around the world. The offices will be manned by experts—experts who understood the global flow of uranium, how the black markets work, where materials might be hiding, how they're tested, their history, the very latest shifts and changes.

"The people in the offices will be there as the all-purpose, one-stop shop for everything that's knowable on the issue," he says. They will be there to bring the best intelligence and insights to those in both the government and private sector. They can even work in the shadows, just as a reporter or investigator might, identifying potential buyers or sellers. If they come across anything illegal, they can report it to local authorities. Mostly, however, they can guide the activities of a state to police its own terrain, while keeping the state tapped into a massive global flow of information.

That's the key, he says. Transparency. Vast information sharing.

"Secrecy is a dying cult," he says. "And it's going to kill us. It's time we stopped treating secrets as some kind of asset, to be traded or hoarded. In this one area, if in no other, everything has to be in sunlight. That's the only way to protect ourselves, or some other country, from having a catastrophic event."

In June, he signs on to do a major speech at the Washington Institute for Near East Policy. It's a top venue, crowded with some of the smartest old pros in Washington. Reporters are told that everything but a short prepared statement will be off the record—still, a large contingent shows up: Associated Press, CNN, NPR, *Washington Post, L.A. Times.*

Rolf has carefully worked on his opening—the prepared remarks—to keep the language fairly bureaucratic, not sensational. To trained ears, which this room is filled with, it's clear he's committing candor.

"While we must continue our work toward improving materials security and reducing levels of nuclear materials stocks, we must also urgently intensify efforts to acquire any materials that may be for sale on the illicit nuclear market," he says. "We must take urgent action to scoop up any nuclear material outside state control before terrorists do."

The suggestion is open-ended—nothing as specific as how this might be accomplished—but the Associated Press and NPR will pick up that general point and send it around the globe.

"It must be a global effort incorporating police, intelligence services, militaries, government agencies and ministries, and citizens across the world," he says. "The effort will require broad and often unprecedented information sharing across every front—between government and private sector, and among foreign partners, including those who were once our adversaries.

And we must take a systems approach that is able to monitor and adjust to fluctuations in all things nuclear across the globe."

That's bureaucratese for the "open-source offices."

The off-the-record Q&A goes on for nearly an hour. There's so much to discuss—A. Q. Khan's recent emergence from four years of house arrest in Pakistan, the destruction of an alleged Syrian nuclear facility, rumors that the United States is preparing to bomb Iran.

Rolf fences with reporters and notable others—such as Efraim Halevy, the former head of Mossad, the Israeli intelligence service—until he finishes with a summation, a long personal plea about the way the world looks now.

"We should understand that we are looking for a needle in a haystack—everywhere we're looking for a [nuclear] facility, a group, a single weapon," he says. "It's not something where we can declare victory when we check off something that some commission says we need to do. We're going to be working on this, increasing our effort, throughout the twenty-first century. We're going to need to bind the effort better together, look at the way the proliferation problem meets the terrorism problem, look at how energy is going to impact on the issues as the world goes increasingly toward nuclear energy," and how "we're going to need to continually adapt our structure to fit that, creating new structures, new ways to look at the problem," and how "we can't submit this to the business as usual model.

"We need to make sure there's cognizance at the highest levels."

Even with the publicity his prepared remarks generated, there's no word, no calls or internal notes, from anyone in the intelligence community or the White House.

He considers this, sitting a few days later in his office in the bowels of the Energy Department, how it probably means he's being iced out for being too public and too frank, for saying there's a very big problem and we need to do more.

That's fine, and expected. Public servants in this era often do their best when they stop caring about whether they'll be fired. The Energy Department's intelligence chief has increasingly embraced that model.

Rolf thinks about his network of open-source offices and wonders "if we have the moral authority at this point for those to be extensions of the U.S. government, or whether people might think they're a front for other activities, other agendas." No, it probably wouldn't work unless it were a huge NGO, a foundation, or an international organization, he says, "but I hope not.

"It would be so great if the U.S. could lead this charge. Wouldn't it?"

Wendy Chamberlin spends a day trying to redesign her website. The Middle East Institute has a large educational arm, where anyone off the street can learn the region's languages or get cultural acclimation, and she's looking to expand those programs. Online is the way to go.

On this late spring afternoon in 2008, after her assistant has left, she finds herself thinking about the big idea, the way to transmit to the world what she considers true American values—values, she feels, that have been twisted in this era by the plans and prerogatives of official power. Over the past months, she's sketched out this idea or that, some combination of the Marshall Plan and the Peace Corps, but different—tailored, somehow, to what's needed now.

And today, like other days, she keeps coming back to the same moment, something that happened in 2005 that changed her.

On that spring day almost exactly three years ago, her helicopter left at dawn from Khartoum, Sudan—the headquarters, in the mid-1990s, of Osama bin Laden—

headed for an enormous refugee camp in Darfur, three hundred miles west.

Chamberlin, then the acting UN High Commissioner for Refugees, had a meeting at the camp with UN officials and representatives of the Sudanese government. Such meetings were always tense. The situation in Darfur was worsening by the day—and it was the kind of crisis she was convinced the world would be seeing more of. The immediate cause was climate change, a rapid rise in temperatures that had turned northern Darfur, the western edge of Sudan that borders Chad, into a wasteland. Most of Sudan's 40 million people were Arabic-speaking Africans, including northern Darfur's African Arab tribes, who were forced by drought to migrate south with their cattle. They began to fight with non-Arab Africans in southern Darfur—a group that had long sought independence—in a conflict that rapidly escalated in 2003, when the Sudanese government began arming northern Darfur's brutal Janjaweed militias. By 2004, as the slaughter—and the displacement of millions—was well under way, Colin Powell called it genocide, "a consistent and widespread pattern of atrocities."

A year later, Chamberlin arrived at an enormous tent city of fifteen thousand refugees. In the few hours before her meeting with government officials, she realized that the entire refugee camp was run by a twenty-seven-year-old American, a young man just four years out of college.

Among the dizzying problems at hand was the matter of how women who had to leave the refugee camp to collect firewood were being raped and murdered by Janjaweed militants. The young man, who worked for an NGO, Refugees International, had negotiated a tenuous truce with the government so that representatives of the African Union—sort of a mini-UN, representing fifty-three African countries—could accompany the women.

"This one kid had to be the liaison to the government, which was hostile—they'd burned all the villages in this region, which had created the camp—while making sure all the food and water actually made it to the people."

In the big tent at midday, the arguments about the attacks on the women raged between Sudanese officials, Chamberlin, and a representative from the UN Human Rights Commission stationed at the camp. The young man was silent.

Afterward, he and Chamberlin stood outside in the 120-degree heat.

"Why didn't you say anything?" she asked.

"If I say anything too strident to the Sudanese officials," he explained, "they'll just kick me out. They'll declare me persona non grata, and then who will do what I do now?"

"I realized," Chamberlin recalls, "that the guy from the UN Human Rights Commission, who was fairly ineffectual, had his role: to wave his finger in the faces of the Sudanese about the women or delayed shipments of food and water. You needed someone with a diplomatic presence, who had some protection.

"But it was the kid—this American kid—who was holding it all together."

Chamberlin remembers standing there, speechless, feeling, she says, the young man's "vulnerability and responsibility. I asked him 'How are you managing this?' "

He didn't say anything for a minute, as though no one had ever asked him this.

"I feel responsible for the lives of these people," he said.

Two years later, sitting in her Washington office, Chamberlin can hear his voice, and see him standing there.

"I'll bet every one of those fifteen thousand people knew that kid, who, without preaching to them or telling them what to do or how to be more like us, was their lifeline. And none of those people he managed to keep alive will ever forget that. They'd met an American."

Today, as she packs up her briefcase, Wendy Chamberlin—who, like so many other characters in this American drama, simply wants to feel the surge of moral energy again—has her program, her big idea.

"I want to multiply that kid by a thousand, by ten thousand, and give him anything he needs."

Usman Khosa is back in America and walking to work again, in his shorts and T-shirt, on a spring day.

What he carries with him as he walks south toward the White House are the echoes of Lahore.

It was an eventful two months back home. His sister was married, and he made a commitment with his girlfriend, Sadia, who'd visited Pakistan to see some relatives and to meet Usman's family. They exchanged vows to be married. At the end of this year, 2008, there will be wedding celebrations for them in Lahore. Reggie, Usman's African American friend from Barnes Richardson, has already written a song that he's eager to perform at the wedding, for Usman, his "brother from another mother."

And Usman made peace with his sister Sadia. The two talked a great deal about the question they'd touched on that night: how each "defines human progress," and whether they could define it together.

Sadia spoke fervently of how she views her faith as progress, how it provides her with a kind of wholeness, not a retreat from life but a fuller engagement with it, where she can find peace within herself and the perfection of the Koran. "It makes me feel free," she told him.

Though these most basic yearnings are shared, Usman knows he will engage with the world, and define progress, in different ways. He will season his faith with a touch of doubt. He will attain what he can, using the tools of reason. He and his future wife will, he imagines, live in America and raise their children here.

Long-buried memories were unearthed by something his sister said to him during their long night of debate, about how he'd "once believed, four or five years ago, that the Koran was absolutely the word of God." Sadia wondered "what happened."

One spring night at a bookstore café, not far from his D.C. apartment, Usman talks about the time his sister was referring to. It was 2001, and he was a college freshman and deeply involved in his faith, in his identity as a Muslim. He not only wore traditional South Asian dress, his *kameez*, but also thought about his life as one of mission. He had company: another boy named Usman, a friend from his private school in Lahore, who'd applied to Connecticut College right after Usman and had also been accepted. They traveled across the New London campus as a sort of duo, the two Usmans. "We felt different from the other kids, but we had each other and understood each other," Usman recalls. "He, my friend Usman, was getting more religious, and we'd challenge each other about who knew more about the Koran, who was purer in faith.

"Even though we were becoming more insular—two Muslims, debating Koran—we felt a kind of kinship with the world, especially Muslims from far away," he says. "That would include those in Afghanistan. Our view was that the Taliban was just practicing a very fervent form of Islam and they should be free to do so, without interference."

He pauses, as it becomes clear in his mind's eye. "I know this may sound strange," he says, "but we were getting ready that spring to go to Afghanistan and join the Taliban, to live that life."

Then 9/11 happened. And then the key moment arrived, in the crowded student center, as everyone watched the Taliban spokesman on the giant flat-screen TV denying responsibility for the attacks. Usman, in his South Asian dress, was terrified as a group of students approached him—kids he thought were going to attack him, but who only wanted to know how he felt at this difficult time.

In the six months that followed, he still felt a strong connection to what was happening across the globe. Once the United States began to bomb Afghanistan, Usman set up a card table in the student union each day to raise money for the "widows and orphans" in Afghanistan and, all told, raised a surprising $5,000. In early 2002, he and his friend Usman traveled to Washington to demonstrate with other Muslims against continuing U.S. involvement in Afghanistan.

"But something had changed, and I realize, looking back, it happened that day in the student union. I was afraid watching the Taliban spokesman, as the only Muslim in a big room of Christians, mostly, and Jews. I thought I knew what they saw when they looked at me. But it turned out I was wrong. Those students who approached me—here I thought they were ready to hit me—and they really wanted to know how I felt, and how they could help me.

"It's hard to be too sure of your assumptions after that," he says, smiling. "I grew up that day. I started a process of change and, I think, growth, that gets me to where I am now. Who knows where I'd be now, or what I'd be doing, if they hadn't reached out to me."

Usman sits quietly for minute, his mind searching across the intervening years. "I guess you could call it the human solution."

That, after all, is what binds our actors, our characters into one idea, one shared purpose.

Each of them, from their distinct origins, points of view, and wildly disparate circumstances—across the axis of faith and reason—is seeking the human solution.

Thomas Jefferson, in calling his newly founded country "the last hope of human liberty in this world," would say that the solution itself is America—the place where certain immutable principles will allow citizens to try to arrive at a common definition of human progress. The key is the attempt, messy and uncertain in its outcome—and often as combative as Ibrahim's collision with Ann Petrila, or Usman's fierce debate with his sister Sadia—but handled according to established moral standards such as honesty and compassion.

If there is one thing that has marked this era and its excesses, it has been a lack of appreciation for the innate unruliness of this process of shared search. The process can't be forced. Those who try usually fail. In some cases, they face charges of acting immorally or illegally.

Yet it seems that after difficult years of confusion and fear, the American model of self-governance is indeed operating effectively, as a system designed to self-correct.

On June 12, the Supreme Court issued its ruling in the combined cases of *Boumediene v. Bush* and *Al Odah v. United States*, dividing along its most common ideological axis, with Justice Anthony Kennedy tipping the balance in favor of the majority. The 5–4 decision restores habeas corpus for the detainees at Guantánamo, affording them the right to challenge their detention in U.S. federal court.

The seventy-page majority opinion, authored by Justice Kennedy, works its way methodically through the complex and, in ways, unprecedented case, ruling that none of the particulars of detention justifies an exemption of habeas, and that "the DTA review procedures are an inadequate substitute for" the Great Writ. When Kennedy explains that the problems inherent in the tribunal process call for a more robust recourse for review and appeal than is currently permitted, he seems to be speaking directly to the detainees themselves. "Even when all parties involved in this process act with diligence and in good faith, there is considerable risk of error in the tribunal's findings of fact," Kennedy writes in the majority opinion. "And given that the consequence of error may be detention of persons for the duration of hostilities that may last a generation or more, this is a risk too significant too ignore."

In his dissent, Justice Antonin Scalia writes that the decision "will make the war harder on us. It will almost certainly cause more Americans to be killed." President Bush, who was in Rome on a European tour when the ruling was announced, said, "It was a deeply divided court, and I strongly agree with those who dissented." He added that he may seek to enact new laws, "so we can safely say to the American people, 'We're doing everything we can to protect you.'"

In his carefully written opinion, Justice Kennedy seems to respond directly to the president. "Security depends upon a sophisticated intelligence apparatus and the ability of our Armed Forces to act and to interdict. There are further considerations, however," Kennedy writes. "Security subsists, too, in fidelity to freedom's first principles."

Candace Gorman, for one, received the news joyously and, in minutes, sent a note to the Supreme Court about the habeas petition for Ghizzawi, which she filed with the Court last year. As she wrote on her blog on the day of the decision, "This afternoon I asked the very nice clerk at the Supreme Court if there was any word on what will happen with that petition. I received a response a few hours ago telling me that the petition is set for the judges conference on June 19th and that an order will probably be entered on June 23rd.

"In my dream of dreams the Supreme Court will rule on the petition and order that Mr. Al-Ghizzawi be set free, but my best guess is that the Court will send the petition back to the District Court for a prompt hearing," she writes. "*Prompt* is not a word that is usually associated with anything Gitmo, but I hope for the best. As Studs Terkel said, 'Hope dies last.'"

It clearly remains alive in America as a troubled, and troubling, decade nears its end.

Those two great genies of wish fulfillment—messianic fervor and technological power—have spotted each other. If they come together, the world, as it is now known, will no longer exist.

Standing firmly in their way is a small community that has been experimenting with the idea that disparate people can, and must, understand one another. That would include those gathered around Ibrahim Frotan, from Denver to Kane, and especially his friend Jillian. It would include Rolf, and his fitful efforts to learn what is knowable and then challenge the idea of secrecy itself by taking his Armageddon Test public. It would include Abdul Hamid al-Ghizzawi, his last friend, Candace Gorman, and their invisible partner

618 • RON SUSKIND

Stephen Abraham, all of whom managed to show that American ideals, and the nation's system of justice, are designed to include, not discard, a powerless prisoner.

What's clear is that the human capacity for survival is asserting itself with some ingenuity. When people are tested—just as they are now by collisions of haves and have-nots, by destructive capacity and faith-driven violence—they often manage to discover saving truths. One, spoken by Benazir Bhutto in her last days, but understood by everyone from Wendy Chamberlin to Ibrahim and Jillian, is about the true way of the world. When the world works, and it often has over the recent centuries, it's because everyone moves forward, in a kind of modest unison. It's not about a giver and receiver, and the obligations accrued therein. Or about who saved whom. It is, as Usman and Sadia agree, about defining human progress together, and making sure everyone advances, even if it's just one step.

In its report about the September 11 attacks, the 9/11 Commission blamed U.S. leaders, and especially the intelligence community, for a "failure of imagination." That failure involved not imagining the enemy's malevolence and ingenuity.

Now it's a lack of imagination again that is America's greatest vulnerability—a lack of imagination about what the nation might yet become. Bin Laden and Zawahiri have their story—an old one of the prince walking from the peak into the valley. It's an ancient story, where those of noble standing show the unwashed multitude that, yes, everyone understands, deep down, how such distinctions of caste and class are hollow, and wrongly define one's place in the world. Nothing changes, though, generation after generation. It's all about the luck of birth, about the hand you're dealt.

But the American story is much greater. It's the one that replaced that old story. It's not about the privileged defending what they have with mighty armies or earnest self-regard.

It's about common people coming to the shores of a vast, challenging place, discovering their truest potential, and re-creating, over and over, a new world.

That's why people across the erupting planet want us to tell that story and help them tell it, too.

It's about the valley rising.

Acknowledgments

I t is with unbounded joy that I'm able to thank the very many people who made this book possible.

There was a team of fine and fierce supporters on this effort, starting from the point of inception. Tim Duggan, HarperCollins executive editor, and the house's publisher, Jonathan Burnham, were incisive enthusiasts from first moments of the project to its finale. Both men have my deep gratitude.

Through twists and turns, some of them startling, Tim was ever my partner, offering wise counsel as an editor and a friend in every conceivable area. He was tireless and ebullient and, at day's end, absolutely fierce in his commitment to the book.

With a complex weave of narrative and news— and some disclosures arriving late in the reporting—

HarperCollins pulled the book together like a battalion in battle. Special thanks goes to production editor John Jusino, copyeditor Jenna Dolan, and Allison Lorentzen, Tim Duggan's assistant and a master of many trades. Trina Hunn, HarperCollins's legal counsel, did an exemplary job with the pressure on, and even called in outside experts to get second and third opinions.

Andrew Wylie, my agent, displayed great wisdom and discretion on this particular adventure from the outset, when it was barely a bright idea, to the raucous finish. He is a good friend.

One of the most fortunate moments in this project came at the start: hiring Greg Jackson as my research assistant. The mix of abilities and profound commitment Greg brought to this effort was nothing short of inspiring. Notable among them was the day in September 2007, when Greg was sent to New York on a project for the book and detained by federal agents in Manhattan. He was interrogated and his notes were confiscated, violations of his First and Fourth Amendment rights. Then next day, he was back at work, with his usual balance of inquisitive energy and sterling judgment, and worked this project night and day, month after month, until its finish.

A journalist is graced with the insights of others and charged with doing justice to those insights in print.

I hope I've done that in this book. There were many sources for this book, quite a few working on deep background. It should be noted that the intelligence sources who are quoted in this book in no way disclosed any classified information. None crossed the line.

I had outside legal assistance on this project from Washington's venerable firm of Covington and Burling, legal support led—as in past years and projects—by one of the town's smartest lawyers, Mark Lynch. Kurt Wimmer, formerly of Covington and now general counsel at Gannett, also jumped in admirably.

Once again, I found a refuge at a crucial time in the book's writing and reporting at Dartmouth College, where I've been a visiting scholar since 2000. I have many friends at Dartmouth—professors who share with me their earned wisdom—among them Ken Yalowitz, a dean of America's diplomatic corps, who was an exemplary adviser and patron. As with other projects, friends stepped up with houses for writing retreats, including the Kiggins clan of Fairlee and my old friends Tom and Melissa Dann.

There are many people who merit thanks, but I'd like to mention an unsung hero, my brother, Len, who displayed great fortitude during some tough times in the past two years and my mother, Shirley, who kept smiling and fighting through some difficult transitions.

My good fortune in this life is due to being blessed with extraordinary sons—Walter, who's nineteen, and Owen, now seventeen. In countless ways, large and small, they've inspired and enlightened me, as has my amazing wife, Cornelia.

This was one tough project. She was a beacon of light, throughout. For reasons beyond expression, this book is dedicated to her.

THE NEW LUXURY IN READING

We hope you enjoyed reading
our new, comfortable print size and found it
an experience you would like to repeat.

Well – you're in luck!

HarperLuxe offers the finest in fiction and
nonfiction books in this same larger print size and
paperback format. Light and easy to read, HarperLuxe
paperbacks are for book lovers who want to see
what they are reading without the strain.

For a full listing of titles and
new releases to come, please visit our website:

www.HarperLuxe.com

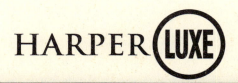

SEEING IS BELIEVING!